Mathematical Circles

Vol III

Mathematical Circles Adieu
Return to Mathematical Circles

Mathematical Circles

Vol III

Howard Eves

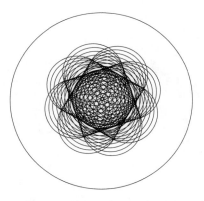

Mathematical Circles Adieu
Return to Mathematical Circles

Published and Distributed by
The Mathematical Association of America

Alexander Woolcott once described maturity as "anecdotage."

Mathematical Circles Adieu was previously published by Prindle, Weber & Schmidt, Incorporated in 1977. *Return to Mathematical Circles* was previously published by Prindle, Weber & Schmidt, Incorporated in 1988.

© *2003 by*
The Mathematical Association of America, Inc.
Library of Congress Catalog Card Number 2002116036

ISBN 0-88385-544-5

Printed in the United States of America

Current Printing (last digit):
10 9 8 7 6 5 4 3 2 1

The Spectrum Series of the Mathematical Association of America was so named to reflect its purpose: to publish a broad range of books including biographies, accessible expositions of old or new mathematical ideas, reprints and revisions of excellent out-of-print books, popular works, and other monographs of high interest that will appeal to a broad range of readers, including students and teachers of mathematics, mathematical amateurs, and researchers.

MAA Service Center
P.O. Box 91112
Washington, DC 20090-1112
800-331-1622   FAX 301-206-9789

PUBLISHER'S NOTE

For many years Howard Eves, famed historian of mathematics and master teacher, collected stories and anecdotes about mathematics and mathematicians and gathered them together in six Mathematical Circles books. Thousands of teachers of mathematics have read these stories and anecdotes for their own enjoyment and used them in the classroom to add spice and entertainment, to introduce a human element, to inspire the student, and to forge some links of cultural history. Through a special arrangement with Professor Eves, the Mathematical Association of America (MAA) is proud to reissue all six of the Mathematical Circles books in this three-volume edition.

In Mathematical Circles, the first two books, were published to acclaim in 1969. They are bound together here as Volume I of the Mathematical Circles Collection. *Mathematical Circles Revisited* and *Mathematical Circles Squared* are bound together as Volume 2 of the Collection, and *Mathematical Circles Adieu* and *Return to Mathematical Circles* as Volume 3.

This three-volume set is a must for all who enjoy the mathematical enterprise, especially those who appreciate the human and cultural aspects of mathematics.

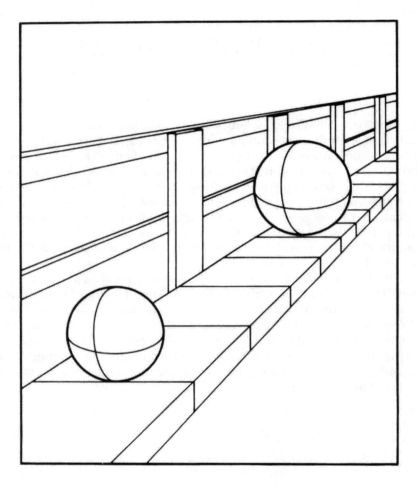

The optical illusion of perspective is mentioned in Item 1456°.

Mathematical Circles
Adieu

Howard Eves

Published and Distributed by
The Mathematical Association of America

TO   THE ADMINISTRATION, FACULTY, AND STUDENTS
of the University of Maine at Machias
with fond memories and warmest wishes

PREFACE

Upon completion of the third trip around the Mathematical Circle (*Mathematical Circles Squared*), I fully intended not to make any more of these circuits. In the intervening five years, however, so many mathematical brethren sent me favorite anecdotes that they hoped would appear in a subsequent ramble, that I finally felt compelled to go around just once more. Herewith is the result, which I insist is my farewell to these pleasant wanderings.

Again I point out that many anecdotes cannot survive a careful examination for veracity; they are to be regarded more as folklore than as truth. Someone once insightfully defined an anecdote as "a revealing account of an incident that never occurred in the life of some famous person."

So, many thanks to all those who have supplied me with material and to those who have written kind and encouraging words about my circular travels. Special thanks go to *The Mathematics Teacher* and to *The American Mathematical Monthly*, to my friend Elwood Ede and to the good folks at the University of Maine at Machias (where I assembled the work while assisting there in the mathematics department), and to Prindle, Weber and Schmidt for their generous and sensitive understanding. Finally, I hope this fourth set of anecdotes will prolong our fond memories of two especially lovable and talented mathematicians, L. J. Mordell and Leo Moser.

Adieu, then, and perhaps at some future time we devotees shall meet in some mathematical Valhalla and enjoy ourselves by swapping more stories of the old days on earth.

HOWARD W. EVES

CONTENTS

CONTENTS

CONTENTS

CONTENTS

QUADRANT TWO

CONTENTS

QUADRANT THREE

CONTENTS

CONTENTS

CONTENTS

QUADRANT FOUR

GEOMETRY

RECREATIONAL MATTERS

L'ENVOI

GEOMETRICAL ILLUSIONS

A combined index for *Mathematical Circles Adieu* and *Return to Mathematical Circles* can be found at the end of this volume.

QUADRANT ONE

From a unique thesis
to Lewis Carroll's fireplace

MATHEMATICS IN EARLY AMERICA

MATHEMATICS was slow getting started in the early American colonies, and that which was done and taught shows a stark poverty compared with the mathematics of the same period in Europe. This was to be expected, for the time and energy of the early colonists were largely consumed by more mundane, immediate, and pressing matters. But, slow and meager as the start was, once the nineteenth century was past, mathematics in our country developed at a prodigious and ever-increasing rate, and today mathematics in the United States is second to that of no other country of the world.

For an excellent brief survey of the history of early American mathematics, see D. E. Smith and Jekuthiel Ginsburg, *A History of Mathematics in America Before 1900* (Carus Mathematical Monograph Number Five). Many of the items of the present section were suggested by this fine little book.

1° *A unique thesis.* The early American colleges were designed primarily for the training of clergy, and consequently followed the English plan of confining their work principally to Latin and Greek, with the result that courses in science, even for the master's degree, were quite trivial. The only mathematical master's thesis offered in the United States before 1700 was: "Is the quadrature of a circle possible?" in which the candidate took the affirmative position. This was in 1693 at Harvard College.

2° *The poverty of early American collegiate mathematics.* The low standards of early American collegiate mathematics are exemplified in the following titles of theses for the bachelor's degree at Yale in 1718.

1. Multiplication by a decimal fraction decreases the value of any given number; division increases it. [Apparently the case of a proper common fraction was not considered as analogous.]
2. What is involved in involution is resolved by evolution.

3

3. Given the base and altitude, the angle at the base cannot be found by the use of a line of sines. [Here we have a right triangle standing on one of its legs as base. It seems that the writer was unaware of the relation $\tan A = \sin A / (1 - \sin^2 A)^{1/2}$.]

4. Trigonometric problems can be solved most accurately by the use of logarithms.

5. The surface of a sphere is four times the area of its largest circle.

6. The angle at the base of the horizontal sundial must agree with the elevation of the pole.

3° *Low-grade work.*   The low grade of mathematical work in the colleges of early America is illustrated by *The Statutes of Columbia College in the City of New York,* printed in 1785. For entrance, a candidate had merely to know the four fundamental operations of arithmetic, along with the rule of three. In the freshman year the student devoted three class sessions a week to vulgar and decimal fractions, extraction of roots, and algebra as far as quadratic equations. In the sophomore year the mathematics class met once a day to study Euclid's *Elements,* spherical trigonometry, conic sections, and the "higher branches of algebra." No further mathematics was required except that taught incidentally in the physics and astronomy courses of the junior year.

4° *Early American theologian–mathematicians.*   The Puritan hope that a Holy Commonwealth would develop in America, and the zeal of such religious leaders as Cotton Mather, Jonathon Edwards, George Whitefield, and John Woolman, naturally left a mark on the early American colleges. In a similar way, Samuel Johnson, the first president of King's College, from which Columbia University later evolved, and Dean (later to become Bishop) George Berkeley tried to revert the people of the colonies back to the Anglican Church. Berkeley later wrote diatribes against the weaknesses in the foundations of Newton's theory of fluxions, and his *Collected Works* deal extensively with mathematics. Dean Berkeley came to Newport, Rhode Island, in 1729 and remained in America for three years.

It is because of this early religious zeal that the first colleges in America, like those in England and Ireland, were established primarily to train the clergy. Thus, up to about 1850, a large proportion of the professors of mathematics in America were clergymen with more interest in theology than in mathematics.

5° *George Berkeley in America.* George Berkeley (1685–1753), who became a well-known Anglican bishop and a severe critic of Newton's theory of fluxions, earlier in life entertained the idea of establishing a college in Bermuda for the education of young colonists and Indians of the American mainland. Though he gathered some high support for his plan, the project finally had to be dropped, but not before he spent three years (1729–1731) in Newport, Rhode Island, where he gave considerable encouragement to higher education in the American colonies. He is still honored at Harvard and at Yale, and Berkeley, California, was named after him.

6° *The Winthrop mathematics library.* Important libraries, from the historical point of view, sometimes settle in unexpected places and rest there often unknown even to most of the local scholars. Perhaps the best evidence of the mathematical material available in America in the mid-eighteenth century is to be found in a collection of over a hundred old texts now residing in the library of Allegheny College in Meadville, Pennsylvania. It was the mathematics library of John Winthrop (1714–1779), one of four Winthrops bearing that name, and a descendent of the first governor of the Massachusetts Bay Colony.

John Winthrop was one of the most learned men in America of his time. When only twenty-four, he was appointed to the Hollis professorship at Harvard, a post that he held for over forty years. Among his first self-appointed tasks at Harvard was to master Newton's *Principia* and to make full use of the astronomical instruments that Thomas Hollis had earlier had sent to Harvard from England. Among the instruments was a telescope which had belonged to Edmund Halley (1656–1742). Winthrop was an astronomer rather than a mathematician, and the original records of his astronomical observations of 1739 are still in the Harvard library. It was in 1740 that he made the first observation in America of a transit of the planet Mercury across the face of the sun, and later, in 1761, he traveled to Newfoundland to observe another such transit, this trip probably being the first purely scientific expedition sent out by an American colony. In 1766 he was elected a Fellow of the Royal Society. Two years later, in 1768, he was awarded the degree LL.D. by the University of Edinburgh and was chosen a member of the American Philosophical Society. He was the first to receive the degree LL.D. from Harvard (in 1773) and he was the principal founder of the American Academy of Arts and Sciences.

Winthrop's above-mentioned mathematics library, now housed at Allegheny College, contains over one hundred books on pure and applied mathematics, and almost all of them are historical classics. The collection contains works by Barrow, Cassini, Cotes, De Lalande, Desaguliers, Descartes, Euclid, Gravesande, Halley, Huygens, Keill, Maclaurin, Maseres, Newton, Oughtred, Ramus, Whiston, and Wolf. Many of these books are very rare, and any one of them would occupy a worthy place in any mathematical collection. One would think this collection would have settled at Harvard. There must be a story as to how Allegheny College acquired the volumes. When the writer of these anecdotes spent a week lecturing at Allegheny College a few years ago, none of the members of the mathematics staff there was aware of the presence of the Winthrop collection.

A collection of over 150 mathematical classics that were in the Harvard library in the first quarter of the eighteenth century was completely consumed by fire in 1764.

7° *Jefferson on Rittenhouse.* David Rittenhouse (1732–1796) was, after John Winthrop, the next noted early American astronomer with an interest in applied mathematics. Starting as a watchmaker, he became a constructer of precision scientific instruments. As an able practical mathematician, he early worked as a surveyor and played a part in the establishment of the Mason-Dixon line. He made a remarkably precise observation of the transit of Venus on June 3, 1769, and very accurately calculated the elements of the future transit of Venus of December 8, 1874. The high quality of his papers in the first four volumes of the *Transactions of the American Philosophical Society* led to his election as president, in 1790, of the Society, succeeding Benjamin Franklin. He was a Fellow of the American Society of Arts and Letters and of the Royal Society of London, and in 1789 was awarded the degree LL.D. by Princeton University. He served as vice-provost at the University of Pennsylvania and was among the first to use spider lines in a telescope. In 1792, George Washington appointed him Director of the Mint, a position he held until the year before his death, and one is reminded of the last years of Isaac Newton.

Thomas Jefferson, always ready to defend his country against critics and to extol the achievements of his countrymen, wrote the following extravagant lines about Rittenhouse and his construction of a planetarium:

> We have supposed Rittenhouse second to no astronomer living; that in genius he must be first because he is self taught. As an artist he has exhibited as great a proof of mechanical genius as the world has ever produced. He has not indeed made a world; but he has by imitation nearer approached its Maker than any man who has lived from Creation to this day.

8° *Burning the calculus.* It was a common custom among mathematics students in American colleges in the late nineteenth century ceremoniously to burn their calculus texts at the close of the sophomore year, thus displaying by a great bonfire their estimate of the value of the subject as then taught. [In this connection, see Item 139° of *Mathematical Circles Revisited.*]

9° *How Charles Gill became a mathematician.* There are a number of names that were prominent in American mathematics in the first half of the nineteenth century but which are now almost completely forgotten—names like Robert Adrain (1775–1843), Robert Patterson (1743–1824), William Rogers (1804–1882), John Farrar (1779–1854), Theodore Strong (1790–1869), Alexander Dallas Bache (1806–1867), and Ferdinand Rudolph Hassler (1770–1843).

Among these earlier and now almost forgotten men was Charles Gill, who had an interesting and unusual entry into mathematics. Gill was born in Yorkshire, England, in 1805, the son of a local shoemaker. When thirteen, Gill went to sea and voyaged to the West Indies. Three years later, when he was only sixteen, the rapid and tragic loss by yellow fever of all the officers of the ship on which he was serving, catapulted him into command of the vessel. His navigational skill brought the ship safely to its destined port, and he became convinced that mathematics was his forte. He accordingly gave up the sea and began a career of teaching mathematics, spending his leisure time studying the subject. He contributed to a number of mathematical journals and was already an established thinker when he came to America in 1830.

In this country, Gill immediately engaged in teaching mathematics in secondary schools, his final position being in the Flushing Institute on Long Island, N.Y. It was while at this school, which grew into a short-lived college, that Gill brought out the *Mathematical Miscellany*, a semi-annual periodical devoted largely to the proposal and solution of fairly difficult and advanced mathematics problems. The journal, though it lasted only through the four years 1836–1839, exerted a considerable influence on American mathematics. Gill himself possessed real skill in solving intricate problems in number theory. Solutions by him of such problems appeared in such popular journals as *The Ladies' Diary, The Gentleman's Mathematical Companion, The Educational Times,* and his own *Mathematical Miscellany.* He was also interested in actuarial mathematics, and has been called "the first actuary in America." Gill is mentioned several times in Dickson's *History of the Theory of Numbers.*

10° *A "paradoxical" relation.* Benjamin Peirce (1809–1880) was generally regarded as the leading mathematician in America of his time. He graduated from Harvard in 1829 when only twenty years old, became a tutor there in 1831, then soon after, in 1833, was awarded the professorship in mathematics and natural science, and still later, in 1842, the professorship in mathematics and astronomy.

The most often told anecdote about Benjamin Peirce concerns an incident that occurred in one of his classes on function theory. Having established the well-known and remarkable relation

$$e^{\pi/2} = {}^i\sqrt{i}$$

on the blackboard, he turned to the class and said, "Gentlemen, that is surely true, it is absolutely paradoxical, we cannot understand it, and we don't know what it means, but we have proved it, and therefore we know it must be the truth."

The above relation is perhaps even more striking in the equivalent form

$$e^{i\pi} = -1,$$

for here we have four of the most important numbers of mathematics, namely e, i, π, and -1, all related by a very simple expression.

11° *A molder of college presidents.*   Among Benjamin Peirce's students was Abbott Lawrence Lowell, who graduated in 1877 at the age of twenty-one with highest honors in mathematics. Lowell's paper on "Surfaces of the second order as treated by quaternions" attests to his continued interest in mathematics. In 1909 Lowell became president of Harvard University, serving in that capacity for twenty-four years.

Thomas Hill (1818–1891) was another and earlier student of Peirce, but Hill did not achieve the level of mathematical recognition later accorded to Lowell. Hill became president of Harvard University in 1863.

12° *The Peirce family.*   The Peirce family was one of the most prominent families in early American mathematics. We have already noted (in Item 10°) that Benjamin Peirce (1809–1880) was a famous professor of mathematics at Harvard and was considered the leading American mathematician of his time. Benjamin Peirce had two sons, James Mills Peirce (1834–1906) and Charles Sanders Peirce (1839–1914). James Mills Peirce served as an assistant professor of mathematics at Harvard for the eight years from 1860 to 1869, and as a professor of mathematics from 1869 to 1906. He was largely an applied mathematician. Charles Sanders Peirce was perhaps the greatest mathematician of the Peirce family, though it is only in more recent times that much of his work has been properly appreciated. He graduated from Harvard in 1859 at the age of twenty, and from the Lawrence Scientific School of Harvard in 1863. For a number of years he served on the United States Coast and Geodetic Survey and achieved a fame for his work on geodesy. In 1880 he lectured at Johns Hopkins University on philosophical logic. The latter part of his life was devoted chiefly to mathematical logic; it is this work that has only rather recently been recognized as truly trail-blazing. Another mathematical member of the Peirce family was Benjamin Osgood Peirce (1854–1914), a second cousin once removed of Benjamin Peirce. This Peirce wrote valuable papers in mathematical physics. His *Mathematical and Physical Papers 1903–1913* was published by the Harvard Press in 1926.

13° *Appreciation from a master.*   George William Hill (1838–1916), the third president of the American Mathematical Society, ranks as one of the better known astronomers who contributed noteworthily to pure mathematics. He graduated from Rutgers University in 1859 and shortly thereafter joined the staff of the *Nautical Almanac.* In astronomy he devoted most of his energies to lunar theory; in mathematics he concentrated on differential equations, determinants, and series, and introduced infinite determinants into mathematics. It was Hill's memoir on infinite determinants that some nine years later led Henri Poincaré to take up the study of their convergence. Other work of Hill lay at the basis of Poincaré's *Les méthodes nouvelles de la méchanique.*

When Professor Robert Simpson Woodward (1849–1924), one of the early prominent supporters of the American Mathematical Society, and its fifth president, introduced Hill to Poincaré on the latter's visit to America, Poincaré's first words as he gripped Hill's hand were, "You are the one man I came to America to see."

14° *A descendent of Benjamin Franklin.*   Benjamin Franklin was an amateur dabbler in magic squares and a few other things of a mathematical nature (see Item 317° of *In Mathematical Circles*).

One might fairly wonder if there ever was a descendent of Benjamin Franklin who might more properly be classified as a mathematician. There was—Alexander Dallas Bache (1806–1867). Though no truly outstanding figure in mathematics, Bache did author several articles on astronomy and surveying in the *Proceedings* and the *Transactions of the American Philosophical Society.* He was, from 1828 to 1841, a professor of natural philosophy and chemistry at the University of Pennsylvania. In 1843 he became superintendent of the United States Coast Survey, and it was here that he made his reputation.

PIERRE DE FERMAT AND RENÉ DESCARTES

WE have noted, in the previous section, the paucity of mathematics and the complete lack of top-flight mathematicians in America in the seventeenth century. In sharp contrast to this sterile picture, mathematics in Europe was thriving, and indeed was experiencing a remarkable period of growth.

It was early in the seventeenth century that Napier revealed his invention of logarithms, Harriot and Oughtred contributed to the notation and codification of algebra, Galileo founded the science of dynamics, and Kepler announced his laws of planetary motion. Later in the century, Desargues and Pascal opened up the field of projective geometry, Fermat fathered modern number theory, Fermat and Descartes gave us analytic geometry, and Fermat, Pascal, and Huygens laid the foundations of the modern theory of mathematical probability. Then, toward the end of the century, after a host of seventeenth-century mathematicians had prepared the way, the epoch-making creation of the calculus was made by Newton and Leibniz. The seventeenth century was truly a remarkable period for mathematics in western Europe.

Of the great European men of mathematics mentioned above, we now tell a few stories about the two Frenchmen, Fermat and Descartes. For many other stories and anecdotes about these men, and about the others mentioned above, the reader may consult our previous three trips around the mathematical circle.

11

15° *The enigma of Fermat's birthdate.*   There is a seemingly reliable report that has come down to us asserting that Fermat was born at Beaumont de Lomagne, near Toulouse, on August 17, 1601. It is known that Fermat died at Castres or Toulouse on January 12, 1665. His tombstone, originally in the church of the Augustines in Toulouse and then later moved to the local museum, gives the above date of death and Fermat's age at death as fifty-seven years. Because of the above conflicting data, Fermat's dates are usually listed as (1601?–1665). Indeed, for various reasons, Fermat's birth year, as given by different writers, ranges from 1590 to 1608.

16° *How did Fermat find the time?*   Fermat did so much top-quality mathematics, in addition to serving in the local parliament of Toulouse, that one naturally wonders where he found the time for all his creative work. It has been suggested that Fermat's position as a *King's councillor* in the parliament of Toulouse is perhaps the key. Unlike many other public servants, a King's councillor was expected to remain aloof from his fellow townsmen and to abstain from most social activities, so that he might less easily be corrupted by bribery and might the better carry out his duties without the stress of partisan influence. In short, Fermat's position gave him plenty of leisure time.

17° *The sixth Fermat number.*   Fermat conjectured that $p_n = 2^{2^n} + 1$ is a prime for all nonnegative integers n. Now for $n = 0$, 1, 2, 3, and 4 we find $p_n = 3, 4, 17, 257$, and $65{,}537$, respectively, all of which are prime numbers. On the other hand, Euler, a century later, showed that $p_5 = 4{,}294{,}967{,}297$ contains 641 as a factor, and therefore is not a prime number. Thus Fermat's conjecture has been proven false. The American lightning calculator, Zerah Colburn, when a mere boy, was once asked if $4{,}294{,}967{,}297$ is prime or not. After a short mental calculation he asserted that it is not, as it has the divisor 641. He was unable to explain the process by which he had reached his amazing correct conclusion.

For further anecdotal information concerning the Fermat numbers, see Item 178° of *In Mathematical Circles*.

18° *Fermat and his method of infinite descent.* In a letter of August, 1659, to Carcavi, Fermat gave the following clear and concise account of his method of infinite descent. We give here a rather free translation of the pertinent part of the letter.

"For a long time I was not able to apply my method to affirmative propositions, because the twist and the trick for getting at them is much more troublesome than that which I use for negative propositions. Thus, when I had to prove that *every prime of the form* $4n + 1$ *is the sum of two squares,* I found myself in a fine torment. But at last a meditation many times repeated gave me the light I lacked, and now affirmative propositions submit to my method, with the aid of certain new principles which necessarily must be adjoined to it. The course of my reasoning in affirmative propositions is such: if an arbitrary chosen prime of the form $4n + 1$ is not a sum of two squares, there will be another of the same nature, less than the one chosen, and next a third still less, and so on. Making an infinite descent in this way we finally arrive at the number 5, the least of all numbers of this kind. It follows that 5 is not the sum of two squares. But it is. Therefore we must infer by a *reductio ad absurdum* that all prime numbers of the form $4n + 1$ are sums of two squares."

Of course, the hard part of applying Fermat's method of infinite descent lies in the very first step, namely proving that if the assumed conjecture is true of any number of the concerned kind, chosen at random, then it will be true of a smaller number of the same concerned kind. In connection with the problem that any prime of the form $4n + 1$ is a sum of two squares, Fermat never recorded for posterity just how he carried out the first step of his method of infinite descent. The result was finally proved by Euler in 1749 after he had struggled, off and on, for seven years to find a proof.

E. T. Bell has pointed out that proofs of many theorems in number theory, like the one above, are so evasive that "it requires more innate intellectual capacity to dispose of the apparently childish thing than it does to grasp the theory of relativity." In the same connection, Gauss wrote: "A great part of its [the higher arithmetic] theories derives an additional charm from the

peculiarity that important propositions, with the impress of simplicity on them, are often easily discovered by induction, and yet
are of so profound a character that we cannot find the demonstrations till after many vain attempts; and even then, when we do
succeed, it is often by some tedious and artificial process, while the
simple methods may long remain concealed."

For a simple illustration of Fermat's method of infinite descent, proving that $\sqrt{2}$ is irrational, see Item 179° of *In Mathematical Circles*.

19° *Descartes versus Fermat.*   Descartes was frequently irrita-
ble, quick-tempered, and acid in his mathematical controversies
with Fermat. Fermat, on the other hand, was always patient, even-
tempered, and unaffectedly courteous. These differing qualities in
the two men are reflected by their avocations—Descartes was a
soldier and Fermat was a jurist.

20° *The Descartes miracle.*   Perhaps the most remarkable
thing about Descartes' life is that it was not brought to a sudden
and premature end by a rapier or a musketball. Descartes spent a
large part of his life soldiering, and though he served much inac-
tive time in army camps, he also on a number of occasions found
himself in the fray of battle. Thus he could easily have lost his life
in 1620, at the young age of 24, when, enlisted under the Elector
of Bavaria, he engaged in the very real fighting of the battle of
Prague. Later he underwent a bloody piece of soldiering, with
distinction, under the Duke of Savoy. Still later, under the King of
France, Descartes' life could have terminated at the siege of La
Rochelle. And if soldiering was not enough, consider the time
when, without his usual complement of bodyguards, he took a boat
for east Frisia and was set upon by a cutthroat crew that sought to
rob him and then throw his body overboard; he whipped out his
sword and compelled the crew to row him back to shore. The
Descartes miracle is that analytic geometry managed to escape the
accidents of battle, sudden death, and murder.

21° *The cavalier.*   Once a half-drunken lout insulted Descartes' lady of the evening. In true cavalier fashion, Descartes went after the sot in stirring D'Artagnan manner and soon flicked the sword out of the fool's hand. He spared the sot's life since he felt it would be too messy to butcher the man in front of a beautiful lady.

22° *Descartes' daughter.*   Though Descartes never married, he did have a daughter by one of his lady friends. The child died early and Descartes was deeply affected.

23° *Overcoming hypochondria.*   Descartes' delicate childhood infected him with hypochondria, and for years he had an oppressive dread of death. Then, at middle age, he came to the conclusion that nature is one's best physician and that the secret of keeping well is to shun the fear of death. Thus he overcame his hypochondria.

24° *Descartes' bones.*   The death of Descartes in Stockholm, while attempting to bring learning to the court of the young and willful Queen Christina of Sweden, has been told in Item 177° of *In Mathematical Circles.* The great philosopher–mathematician was entombed in Sweden and efforts to have his remains transported to France failed. Then, seventeen years after Descartes' death, when Christina had long cast off her crown and her faith, the bones of Descartes, except for those of his right hand, were returned to France and re-entombed in Paris at what is now the Panthéon. The bones of the right hand were secured, as a souvenir, by the French Treasurer-General for his skill in engineering the transportation.

Commenting on the return of Descartes' remains to his native land, Jacobi remarked, "It is often more convenient to possess the ashes of great men than to possess the men themselves during their lifetime."

SOME PRE-NINETEENTH-CENTURY MATHEMATICIANS

HERE are a few random anecdotes, not already told in our earlier trips around the mathematical circle, ranging from Greek antiquity up through the eighteenth century.

25° *A voluptuous moment.* Everyone knows the story of Hobbes's first contact with Euclid: opening the book, by chance, at the theorem of Pythagoras, he exclaimed, "By God, this is impossible," and proceeded to read the proofs backwards until, reaching the axioms, he became convinced. No one can doubt that this was for him a voluptuous moment, unsullied by the thought of the utility of geometry in measuring fields.

—BERTRAND RUSSELL
In Praise of Idleness. London: George Allen and Unwin, Ltd.,
1935.

26° *Faith versus reason.* . . . There can be no doubt about faith and not reason being the *ultima ratio.*

Even Euclid, who has laid himself as little open to the charge of credulity as any writer who ever lived, cannot get beyond this. He has no demonstrable first premise. He requires postulates and axioms which transcend demonstration, and without which he can do nothing. His superstructure indeed is demonstration, but his ground is faith. Nor again can he get further than telling a man he is a fool if he persists in differing from him. He says "which is absurd," and declines to discuss the matter further. Faith and authority, therefore, prove to be as necessary for him as for anyone else.

—SAMUEL BUTLER
From Chapter Sixty-five of *The Way of All Flesh.*

27° *The Gresham chair in geometry.* The earliest professorship of mathematics established in Great Britain was a chair in geometry founded by Sir Thomas Gresham (1519?–1579) at

16

Gresham College in London. Henry Briggs, who was the first occupant of the Savilian chair at Oxford, also had the honor of being the first to occupy the Gresham chair in geometry.

28° *A tenuous connection with mathematics.* Sir Thomas Gresham, known in mathematics as the founder of the first professorship of mathematics in Great Britain, namely the Gresham chair in geometry at Gresham College of London, also founded England's Royal Exchange. Accordingly, when a weather vane was constructed for Faneuil Hall, the famous trading hall of colonial Boston, the designer of the vane fashioned it in the form of a copper grasshopper, for this familiar insect appears on the crest of Sir Thomas Gresham. The designer was Shem Drowne, a metalsmith of Kittery, Maine, who late in the seventeenth century moved to Boston.

The famous grasshopper weather vane atop of Faneuil Hall was noticed missing on January 5, 1974, and there was much speculation about the possibility of thieves having acquired the valuable vane by means of a swooping helicopter. The vane, valued in the hundreds of thousands of dollars, was later prosaically rediscovered wrapped in rags where would-be thieves had left it in the Faneuil Hall belfry.

Four of Shem Drowne's weather vanes are still surviving, and three of these are still in use: the North Church banner (that capped the Old North Church at the time signal lanterns were hung there to warn that the British were moving toward Concord by sea); a rooster vane on a church in Cambridge; and the priceless grasshopper vane of Faneuil Hall.

29° *Gresham's law.* Gresham's law, named after Sir Thomas Gresham, founder of the Royal Exchange of London, states that "bad money tends to drive good money out of circulation." Gresham stated the principle as follows: "When by legal enactment a government assigns the same nominal value to two or more forms of circulating medium whose intrinsic values differ, payments will always, as far as possible, be made in that medium of which the cost of production is least, the more valuable medium tending to disappear from circulation."

Thus, in America in 1896, gold coins were hoarded because of the chance that the government might make silver coins and proclaim them to be legally of the same value as the gold coins. Again, in Europe just after World War I, so much paper money was made that the intrinsically more valuable metal money all but disappeared. Indeed, when paper money was made in America, possessers of silver dollars hoarded them as more valuable than the legally equivalent dollar bills.

As early as the thirteenth century, dishonest dealers would shave the edges of gold coins, and those who received these lighter pieces usually passed them along as quickly as possible, retaining the full-weighted coins as long as they could.

30° *The Savilian and Lucasian professorships.*   Many distinguished British mathematicians have held either a Savilian professorship at Oxford or a Lucasian professorship at Cambridge.

Sir Henry Savile was one time warden at Merton College at Oxford, later provost of Eton, and a lecturer on Euclid at Oxford. In 1619, he founded professorial chairs at Oxford, one in geometry and one in astronomy. Henry Briggs (of logarithm fame) was the first occupant of the Savilian chair of geometry at Oxford. John Wallis, Edmund Halley, and Sir Christopher Wren are other seventeenth-century incumbents of Savilian professorships.

Henry Lucas, who represented Cambridge in parliament in 1639–1640, willed the university resources for the founding in 1663 of the professorship that bears his name. Isaac Barrow was elected the first occupant of this chair in 1664, and six years later was succeeded by Isaac Newton.

31° *Galileo's power.*   The two things most universally desired are power and admiration. Ignorant men can, as a rule, only achieve either by brutal means, involving the acquisition of physical mastery. Culture gives a man less harmful forms of power and more deserving ways of making himself admired. Galileo did more than any monarch has done to change the world, and his power immeasurably exceeded that of his persecutors. He had therefore no need to aim at becoming a persecutor in his turn.

—BERTRAND RUSSELL

In Praise of Idleness. London: George Allen and Unwin, Ltd., 1935.

32° *A clerihew.*   A *clerihew* is a form of light verse, akin to the limerick, that became popular in England. It is named after its inventor, Edmund Clerihew Bentley, well known as the author of *Trent's Last Case,* and the friend to whom G. K. Chesterton dedicated his entertaining fantasy, *The Man Who Was Thursday.* Bentley's best known clerihew concerns the mathematician–architect Sir Christopher Wren (1632–1723):

> Sir Christopher Wren
> Said, "I am going to dine with some men.
> If anybody calls
> Say I'm designing St. Paul's."

As another example of a mathematical clerihew, there is the following, given by J. C. W. de La Bere in the December 1974 issue of the *Australian Mathematical Society Gazette*:

> Archimedes of Syracuse,
> To get into the news,
> Called out "Eureka"
> And became the first streaker.

33° *Euler and the seat of honor.*   In 1783 Czarina Katherine II appointed the talented and celebrated Princess Daschkoff to the directorship of the Imperial Academy of Sciences in Petersburg. Since women were seldom so highly honored in those days, the appointment received wide publicity.

The Princess decided to commence her directorship with a short address to the assembled savants of the Academy. She invited Euler, then elderly and blind but the most respected scientist in Russia, to be her special guest of honor. Euler, with a son and a grandson, accompanied the Princess to the Academy in her personal coach. In her brief address, the Princess stressed her high respect for the sciences and said she hoped that Euler's presence would serve as a source of inspiration to the entire Academy. She then sat down, intending Euler to occupy the chair of honor next to her, but before Euler could be led to the seat, Professor Schtelinn dropped himself into the honored place. Some years ago, Schtelinn had been recognized by Peter III, had served as a councilor of state, and had held the rank of Major-General, and he now felt that he deserved first place in the Academy body.

When Princess Daschkoff saw Schtelinn settling himself next to her, she turned to Euler and said, "Please be seated anywhere, and the chair you choose will naturally be the seat of honor." This act charmed Euler and all present—except the arrogant Professor Schtelinn.

34° *A curious error.*   Leonhard Euler assembled a physics textbook entitled *Lettres à une Princesse d'Allemagne sur divers sujets de physique et de philosophie* [*Letters to a German Princess on various topics of physics and philosophy*]. The letters referred to were originally addressed to Princess Phillipine von Schwedt, niece of Frederick the Great. During the Seven Years War (1756–1763), Princess Phillipine and the entire Berlin court sojourned in Magdeburg. It was during this period that Euler tutored the Princess by letters written from his home in Berlin.

In the letter dated August 27, 1760, Euler enters into an explanation to his royal student of how a surveyor uses a level. He asks the Princess to imagine a straight line drawn from her quarters in Magdeburg to his home in Berlin, and then asks whether this line would be horizontal. He goes on to answer his own question in the negative, stating that the Berlin end of the line would be higher than the Magdeburg end. He explains that Berlin lies on the Spree and Magdeburg on the Elbe, and since the Spree flows into the Havel, and the Havel flows into the Elbe, Magdeburg must be nearer sea level than Berlin.

Now, actually, the Elbe at Magdeburg is 41 meters above sea level while the Spree at Berlin is only 33 meters above sea level, a relation exactly opposite to that deduced by Euler. Since the Spree does indeed flow into the Havel, and the Havel into the Elbe, wherein lies the flaw in Euler's reasoning? Consultation of a map (see our accompanying rough sketch in Figure 1) shows that, though the Havel flows into the Elbe, it does so far below Magdeburg. It is difficult to comprehend how Euler failed to take this fact into account, and possible explanations of the great mathematician's carelessness have been offered.

Some feel that Euler's error was simply that he assumed Magdeburg was near Wittenberge, which lies just below the junction of the Havel with the Elbe. If the imagined line had run from Wittenberge to Berlin, Euler's conclusion would have been correct. The fact is, however, that Magdeburg lies some 55 miles above the junction of the two rivers.

Elwood Ede, formerly of the University of Maine mathematics faculty, suggests a different and perhaps more plausible explanation. He feels that it is very likely Euler merely misjudged the direction of flow of the Elbe, assuming it runs from Wittenberge to Magdeburg instead of from Magdeburg to Wittenberge. Such a misjudgment could easily have occurred, says Ede, if Euler used a map, perhaps cropped, instead of personal observation. In this light, Ede contends, the story becomes somewhat trite and the error, though it exists, is overemphasized.

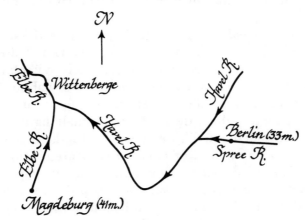

FIGURE 1

35° *Poisson's false start in life.* Siméon Denis Poisson (1781–1840), who became one of the great applied mathematicians of his time, was first destined, much against his own wishes, to be a doctor. The education was undertaken by his uncle, who started the boy off with pricking veins in cabbage leaves with a lancet. When he had perfected himself in this, he was graduated to putting on blisters. But in almost the first case in which he did this by himself, his patient died within a few hours. Although the doctors of the place assured him that "the event was a very common one," he vowed to have nothing more to do with the profession.

36° *The good life.*   Poisson once remarked: "Life is good for only two things, discovering mathematics and teaching mathematics."

CARL FRIEDRICH GAUSS

A N astonishing number of anecdotes have come down to us concerning the great German mathematician Carl Friedrich Gauss (1777–1855). We have already, in our earlier trips around the mathematical circle, told over forty of them. Here are nine more.

37° *An unintended pun.*   Wilhelmine Gauss, the second child of Carl Friedrich Gauss and his first wife, was named after Gauss's close friend, the noted astronomer Heinrich Wilhelm Matthias Olbers of Bremen. Gauss has told a pretty story about Minna, as Wilhelmine was always called, when she was five years old (1813). One evening he told the little girl that the small, light, rosy clouds in the sky are called "cirri", or "fleecy", clouds. A few evenings later, when the same kind of clouds reappeared, Minna tried to detain her father by saying, "Stay a little longer, Papa, the sky is so *belämmert* this evening." [*Belämmert* means "belambed" (suggested to Minna by the "fleecy" clouds), but sounds very much like *belemmert,* which means "befouled".]

22

38° *Minna gets married.* In 1830 Minna Gauss married Heinrich Ewald, the young professor of theology and Oriental languages at Göttingen University. Ewald was incredibly absent-minded in nonacademic matters, a trait that manifested itself in connection with his engagement and marriage to Minna. Friends had determined that the bachelor Ewald should, for his own good, get married. Ewald obligingly agreed, leaving the selection of the lady to the friends, who soon settled on Gauss's two daughters Minna and Theresa as the best possibilities. They decided to leave the choice between these two young ladies to Ewald himself. Accordingly, a tea was arranged at the Gauss home at which the two daughters were to take turns pouring. Ewald was instructed to indicate his choice by accepting tea from the daughter he preferred, a procedure to which he readily acquiesced. On the way home after the tea, the matchmakers congratulated Ewald and explained that Gauss was quite agreeable to Ewald wooing Minna, from whom he had accepted tea. Ewald didn't know what his friends were talking about, for he had completely forgotten, as though he had never heard about it, the whole plot and even the purpose of the tea. But he did woo Minna and married her on September 15, 1830. Before the ceremony on the wedding day, Ewald disappeared, and, after a frantic search, was finally found atop a ladder in front of his bookshelves; he had forgotten all about the impending affair. Curiously enough, the marriage turned out to be a very happy one for both Ewald and Minna, and Gauss was highly pleased to have his colleague Ewald as a son-in-law.

39° *Ewald babysits.* Minna was of frail health and died of tuberculosis on August 12, 1840, when she was only thirty-two years of age. Some years later, her husband Ewald remarried and in 1850 a daughter, whom Gauss regarded almost as his own grandchild, was born of this second marriage. One day, when Ewald's second wife had to go shopping, the baby was left in care of her father, who was very engrossed in some Arabic grammar. When the wife returned home, she could not find the baby, and Ewald was equally perplexed over the child's disappearance. Then a small whimpering revealed that the baby was in a closed dresser drawer, where Ewald apparently had placed it, perhaps thinking it was the safest place for the child during his wife's absence.

One wonders about some of Gauss's colleagues. For two stories concerning the amazing logic of George Julius Ribbentrop, professor of law at Göttingen University, see Items 190° and 191° of *Mathematical Circles Squared*. Nor was the Prince himself free of occupational maladies—see Item 193° of *Mathematical Circles Squared*.

40° *Gauss and Ceres.*   There is a collection of over 1500 minor planets, called asteroids, almost all of whose orbits lie between those of Mars and Jupiter. The largest, and first discovered, of these is Ceres, with a diameter of about 470 miles, or about one-fifth that of the moon; the smallest asteroids have diameters under a mile.

Ceres was discovered in Palermo on New Year's day of 1801 by the Italian astronomer Giuseppe Piazzi (1746–1826), and a few observations of its positions were made during the short six weeks it was in view before it disappeared in the western sky. Astronomers awaited its reappearance in the eastern sky, but in spite of an extensive search, they were unable to find it. From the meager observations that had been made of Ceres, Gauss managed to compute the entire orbit of the little planet, with the result that on New Year's Day of 1802, Heinrich Wilhelm Matthias Olbers (1758–1840), German astronomer and physician of Bremen, was able to refind the elusive body. This mathematical calculation probably brought more popular fame to Gauss than anything else he did during his lifetime; it also commenced a long friendship between Gauss and Olbers.

It is interesting that the three principals involved in the above story were subsequently honored in celestial nomenclature, for the 1000th, 1001st, and 1002nd asteroids discovered were named Piazzia, Gaussia, and Olbersia, respectively.

41° *A priceless souvenir.*   It was at the age of nineteen that Gauss discovered that a regular polygon of seventeen sides can be constructed with straightedge and compasses, and it was this discovery, Gauss once declared, that finally settled for him that mathematics would be the field of study upon which he would spend his life and talents. The slate upon which Gauss carried out his work on the construction of regular polygons with Euclidean tools was given by him, in a moment of ecstasy and affection, to his deep personal college-day friend Wolfgang Bolyai. For the rest of his life, Bolyai kept and treasured the slate as a much-revered souvenir of his friendship with and admiration for the great mathematician. One wonders where that historic slate is today.

For the disposition of another rare and precious Gauss souvenir, see Item 202° of *Mathematical Circles Squared.*

42° *An unusual honor.*   The three-masted schooner used by the 1903 German expedition to the South Pole, headed by Professor Erich von Drygalski of Munich, was named the *Gauss,* and a volcanic peak discovered by the expedition was named the *Gaussberg.*

Some years later the schooner was bought by the United States government, its name was changed, and it was put into service along the west coast of North America.

43° *Gauss's pocket watch.*   Gauss ceased breathing at 1:05 A.M. on February 23, 1855. His pocket watch, which he had carried with him most of his life, ceased ticking at almost exactly the same time.

44° *Gauss's brains.*   With the permission of the family, Gauss's skull and brains were carefully weighed and measured after his death. Though the brains exhibited unusually deep convolutions, they were not particularly weighty or bulky. The brains of Byron, Cuvier, and Schiller were heavier, Dante's lighter. Robert Gauss, a grandson of the great mathematician and a brilliant Colorado newspaper editor, left instructions for his brains to be weighed when he died; they were three ounces heavier than those of his famous grandfather. Gauss's brains, and also those of Dirichlet, are preserved in the department of physiology at Göttingen University.

45° *The Prince of Mathematicians.* On January 10, 1855, by orders of the King of Hanover, the official court sculptor, Christian Heinrich Hesemann, arrived at the Gauss home to start work on a medallion of the great mathematician, who was seriously ill and not expected to live much longer. Hesemann suddenly died, on May 29, 1856, before the work was finished, and the medallion was completed by C. Dopmeyer, another sculptor of Hanover. The medallion was later placed, as a plaque, on Gauss's tombstone in the St. Albans Cemetery in Göttingen. Right after the death of Gauss on February 23, 1855, the King of Hanover ordered that a commemorative medal be prepared in honor of Gauss. This seventy-millimeter medal was in time (1877) completed by the well-known sculptor and medalist, Friedrich Brehmer, of Hanover, and was based on the earlier medallion. On it appears the inscription:

> Georgius V. rex Hannoverge
> Mathematicorum principi
> [George V, King of Hanover
> to the Prince of Mathematicians]

Ever since, Gauss has been known as "the Prince of Mathematicians."

SCHELLBACH AND GRASSMANN

KARL Schellbach (1804–1892) [Cajori says he was born in 1809] and Hermann Günther Grassmann (1809–1877) each spent a large part of his life teaching mathematics to younger students, and each possessed deep religious feelings. Schellbach, though enormously admired and of considerable fame in his day, is little known in mathematical circles now, for his forte was the teaching of mathematics rather than the creation of it. On the other hand, Grassmann, though unnoted and obscure in his day, is widely known in mathematical circles now, for his forte was the creation of mathematics rather than the teaching of it. It is thus that history awards its laurels.

A number of pretty stories about Schellbach and Grassmann have been preserved for us by Dr. W. Ahrens in his engaging little booklet *Mathematiker-Anekdoten*. We revive some of these stories here, adapting free translations from Ahrens made by Mr. Elwood Ede.

46° *Preaching the gospel of mathematics.* Karl Schellbach, the romantic poet and famous mathematics professor at the Friedrich Wilhelm Gymnasium in Berlin, considered the teaching of mathematics to be a religious vocation, and he believed that mathematicians were priests who should expose as many people as possible to the realms of mathematical blessedness and glory. He felt that both the gifted and the intellectually poor should take part in this kingdom of heaven. The Minister of Culture, von Bethmann-Hollweg, grandfather of the fifth Reichskanzler, once commented that Schellbach's teaching was "an inspired hymn of mathematics." In an effort to expose inexperienced teachers to the mathematical instruction of Schellbach, the ministry introduced a special seminar to be led by the renowned professor. A great number of young mathematicians were introduced to the art of teaching by "old Schellbach" at these seminars. Many of these young men went on to become truly outstanding secondary school teachers.

[In mathematics, Schellbach is today chiefly remembered for his elegant analytical solution of Steiner's generalization of Malfatti's problem, wherein, given three conics on a conicoid, one is to determine three others which shall touch two of the given and two of the required conics.]

47° *Schellbach previsioned.* Novalis (1772–1801) preceded Schellbach in proclaiming mathematics to be a religion, and mathematicians to be the blessed priests of the religion. He once commented: "The life of God is mathematics; all divine ambassadors must be mathematicians. Pure mathematics is religion. Mathematicians are the only blessed people."

On mathematics as a branch of theology, see F. De Sua's discerning remark at the conclusion of Item 294° of *In Mathematical Circles*.

For some further, and sometimes penetrating, remarks on mathematics made by Novalis, see Item 223° of *Mathematical Circles Revisited.*

48° *The danger of mathematics.*   Prinz Kraft zu Hohlenlohe-Ingelfingen (1827–1892), a Prussian general of artillery and author of several books on military science, studied under Karl Schellbach. Observing Professor Schellbach, Prinz Kraft said, "Mathematics is indeed dangerous in that it absorbs students to such a degree that it dulls their senses to everything else."

49° *A homework assignment.*   To punish a student for a minor infraction, Schellbach gave the student a homework assignment which on completion the lad was to bring to the professor's home. The student did this, but found the professor was not available. He therefore gave the assignment to one of the professor's daughters. The next day in class the following dialogue took place:

Schellbach: "Why didn't you appear?"
Student: "I was there, Professor, and delivered the work. Someone said you were indisposed."
Schellbach: "To whom did you give the work?"
Student: "To one of your daughters, Professor."
Schellbach: "Which one?"
Student: "I don't know her name, Professor."
Schellbach: "Was she pretty?"
Student: "Yes, Professor."
Schellbach: "Well, then, that was Florence. Very good."

50° *Beauty and truth.*   Georg Wilhelm Friedrich Hegel (1770–1831), the eminent German philosopher, said, "Who does not know the works of the ancients dies without knowing *beauty.*" Karl Schellbach responded, "Who does not know the works of the mathematicians and scientists dies without knowing *truth.*"

51° *Hegel speaks out.*   The discovery [of the small planet Ceres] was made by G. Piazzi of Palermo; and it was the more

interesting as its announcement occurred simultaneously with a
publication by Hegel in which he severely criticized astronomers
for not paying more attention to philosophy, a science, said he,
which would at once have shown them that there could not possibly be more than seven planets, and a study of which would therefore have prevented an absurd waste of time in looking for what
in the nature of things could never be found.

—W. W. R. BALL
A Short History of Mathematics, p. 448.

52° *A man of wide interests.*   The mathematician Hermann
Günther Grassmann of Stettin, Germany, had very broad intellectual interests. He was not only a teacher of mathematics, but of
religion, physics, chemistry, German, Latin, history, and geography. He wrote on physics and composed school texts for the study
of German, Latin, and mathematics. He was a copublisher of a
political weekly in the stormy years of 1848 and 1849. He was
interested in music and in the 1860s he was an opera critic for a
daily newspaper. He prepared a philological treatise on German
plants, edited a missionary paper, investigated phonetic laws,
wrote a dictionary to the Rig-Veda and translated the Rig-Veda in
verse, harmonized folk songs in three voices, composed his great
treatise *Ausdehnungslehre,* and brought up nine of his eleven children. His father was a teacher of mathematics and physics in the
gymnasium at Stettin. His son Hermann Grassmann (b. 1859) also
became a mathematician. The father wrote two books on mathematics, and the son wrote a treatise on projective geometry.

It was at the age of fifty-three that, through general lack of
appreciation of his remarkable work, *Ausdehnungslehre,* Grassmann
gave up mathematics and directed the bulk of his intellectual energies to the study of Sanskrit, achieving in philology splendid results that were much better appreciated than his contributions to
mathematics. Grassmann spent his entire life in his native city of
Stettin, except for the years from 1834 to 1836, when he taught
mathematics in an industrial school in Berlin, having succeeded
Jacob Steiner to the post.

53° *A good tactic.*   A special feature of Grassmann's classes was the so-called *Plauderminuten* (literally, "minutes for idle chatter"). Whenever he and the conscientious students in the class had worked steadily and intensely for some time on mathematics, Grassmann would call a halt and give his students and himself a short pause for free conversation. Grassmann would bring this about simply by sitting on the rostrum and calling, "Plauderminute."

54° *A doubtful tactic.*   One day in class, when some troublemakers were being unruly and were disturbing the entire class, Grassmann's pleas and admonitions went unheeded. Instead of taking disciplinary action, the professor descended from the podium and began praying long and loud to God, asking that He not take the ill behavior of the students into His account of their sins. He prayed that the students should seek to better themselves through His Grace, and so on. It is unlikely that this action yielded more than a temporary solution to the disciplinary problem.

55° *Forgiveness and a kiss.*   It seems that one day a student of the Obertertia sorely vexed Grassmann during a course in religious thought. As a result Grassmann was much too angry to give his customary prayer at the close of the hour. The student, noting the omission, became remorseful, and after class went repentantly to Grassmann and asked to be forgiven. Naturally the kind teacher forgave the boy and kissed him.

SEVEN MATHEMATICIANS AND A POET

THE seven mathematicians János Bolyai, Peter Gustav Lejeune Dirichlet, James Joseph Sylvester, Arthur Cayley, Carl Gustav Joseph Jacobi, William Thomson (Lord Kelvin), Augustus De Morgan, and the poet Henry Wadsworth Longfellow were all born in the first quarter of the nineteenth century.

56° *A misjudgment.*   "You should detest it just as much as any evil practice; it can deprive you of all your leisure, your health, your rest, and the whole happiness of your life. This abysmal darkness might perhaps devour a thousand Newtons; it will never be light on earth . . . ."

[From a letter of 1820 written by Farkas Bolyai (1775–1856) to his son János Bolyai (1802–1860), attempting to discourage the latter from investigating the parallel postulate.]

57° *Dirichlet as a teacher.*   Peter Gustav Lejeune Dirichlet (1805–1859) has been described as possessing a noble, sincere, human, and modest disposition, but he seemed unable to communicate with young minds. When a schoolmate expressed envy because Dirichlet's son could always receive help from his gifted father, the son gave this lamentable but memorable reply: "Oh! My father doesn't know the little things anymore." Dirichlet's waggish nephew, Sebastian Hensel, wrote in his memoirs that the mathematics instruction he received in his sixth and seventh years at the gymnasium from his uncle was the most dreadful experience of his life.

—ELWOOD EDE

(freely translated from W. Ahrens, *Mathematiker-Anekdoten*)

58° *Dirichlet as a correspondent.*   Dirichlet was polite and kind but very reserved, and he seldom committed himself to written communication. Once, when a member of the family received a letter from him, the letter was kept as "an extremely rare document."

When Dirichlet's first child arrived, Dirichlet failed to write of the event to his father-in-law, who was living in London at the time. The father-in-law, when he finally found out, commented that he thought Dirichlet "should have at least been able to write $2 + 1 = 3$." This witty father-in-law was none other than Abraham Mendelssohn, a son of the philosopher Moses Mendelssohn, and father of the composer Felix Mendelssohn.

—ELWOOD EDE

(freely translated from W. Ahrens, *Mathematiker-Anekdoten*)

59° *A doctored notice.* Professor G. Arendt attended, in the summer of 1854 at the University of Berlin, a series of lectures by Dirichlet which Arendt published fifty years later in 1904. The following anecdote in connection with those lectures comes from the professor's letters.

It seems that, during the semester, Dirichlet was forced to postpone the lecture on "Definite integrals" because he became ill. He informed his students by posting the following notice, written in his tiny script, on the door of Room 17:

Because of illness I cannot lecture today
Dirichlet

Dirichlet's absence continued for almost a week while the notice remained on the door. Finally one of the students doctored up the message so that it took the form of a definite integral as follows:

$$\int_{Easter}^{Michaelmas} (\text{Because of illness I cannot lecture today})\ d(1\ \text{Frdor})$$

Dirichlet

It must be realized that 1 Friedrichsd'or was the customary honorarium for a semester lecture, and that Michaelmas is the feast of St. Michael celebrated on September 29.

—ELWOOD EDE
(freely translated from W. Ahrens, *Mathematiker-Anekdoten*)
[One is reminded of the doctoring of a similar notice posted one day by Professor William Thomson (Lord Kelvin) on his lecture-room door. See Item 348° of *Mathematical Circles Revisited.*]

60° *Sylvester on graph theory.* J. J. Sylvester (1814–1897) explained the abstract nature of graph theory as follows: "The theory of ramification is one of pure colligation, for it takes no account of magnitude or position; geometrical lines are used, but these have no more real bearing on the matter than those employed in genealogical tables have in explaining the laws of procreation."—LEO MOSER

61° *Absent-mindedness.*   The following anecdote, although clearly not quite factual, illustrates the reputation for absent-mindedness that Sylvester enjoyed.

Upon arrival from England at Johns Hopkins University, Sylvester was asked to give a major address at the university. He had prepared a manuscript for the address, but after he was introduced and called to the podium he could not locate his manuscript. He asked that the lecture be postponed. On return to his office he made a thorough search but still could not find his manuscript. He decided that he must have left it in England, so he returned overseas to look for it. When he arrived at his home he again made a diligent search for the lost manuscript and at long last he found it—in his pocket.—LEO MOSER

62° *Acquiring a new surname.*   James Joseph, the youngest of a number of children born to Abraham Joseph, was born in London on September 3, 1814. The eldest son eventually migrated to the United States, where, for some reason not now known, he took on the surname Sylvester. The rest of the family soon followed the example, and ever after James Joseph was known as James Joseph Sylvester or, more briefly, as J. J. Sylvester.

63° *Sylvester's first contact with the United States.*   Sylvester's American brother, who was an actuary, suggested to the Directors of the Lotteries Contractors of the United States that they submit a difficult problem in arrangements that was bothering them to his younger brother James Joseph, then only sixteen years old. James's complete and satisfying solution of the problem caused the Directors to award the young mathematician a prize of five hundred dollars.

64° *Sylvester and music.*   In addition to a deep interest in poetry, Sylvester was also interested in music and was, in fact, an accomplished amateur. At one time he took singing lessons from the famous French composer Charles François Gounod and on occasions entertained at workingmen's gatherings with his songs.

It is said that he was prouder of his high C in singing than he was of his invariant theory in mathematics.

65° *Unconscious plagiarism.* Plagiarism in mathematics is not always deliberate but sometimes occurs quite unconsciously. Consider the following cases, cited by J. J. Sylvester of himself, in *The Philosophical Transactions of the Royal Society* (1864), p. 642.

"Professor Cayley has since informed me that the theorem about whose origin I was in doubt, will be found in Schläfli's *De Eliminatione.* This is not the first unconscious plagiarism I have been guilty of towards this eminent man whose friendship I am proud to claim. A more glaring case occurs in a note by me in the *Comptes Rendus,* on the twenty-seven straight lines of cubic surfaces, where I believe I have followed (like one walking in his sleep), down to the very nomenclature and notation, the substance of a portion of a paper inserted by Schläfli in the *Mathematical Journal,* which bears my name as one of the editors upon the face."

66° *The only instance on record when Cayley lost his temper.* Arthur Cayley (1821–1895) was a singularly serene, even-tempered, and always unruffled individual. Only once is it known that he lost his calmness. It seems that one day when he and his friend Sylvester were discussing the theory of invariants in Cayley's law rooms, an office boy came in and handed Cayley a sheaf of legal papers for perusal. A glance at this pile of dismal work coming in the middle of his interesting mathematical discussion with Sylvester caused Cayley to flare up. He seized the batch of papers from the interrupting boy and, with an exclamation of disgust, hurled the material on the floor, and then went on talking mathematics with Sylvester.

67° *Cayley's style.* Shortly after Cayley died in 1895, A. R. Forsyth wrote an "appreciation" of the great mathematician and his work in *The Proceedings of the London Royal Society,* Vol. 58 (1895). In this paper Forsyth, writing of Cayley's style, says:

"When Cayley had reached his most advanced generalizations he proceeded to establish them directly by some method or other, though he seldom gave the clue by which they had first been obtained: a proceeding which does not tend to make his papers easy reading. . . .

"His literary style is direct, simple and clear. His legal training had an influence, not merely upon his mode of arrangement but also upon his expression; the result is that his papers are severe and present a curious contrast to the luxuriant enthusiasm which pervades so many of Sylvester's papers."

68° *More on the Jacobi brothers.*   Item 344° of *In Mathematical Circles* reports the confusion that existed between the mathematician Carl Gustav Joseph Jacobi (1804–1851) and his brother Moritz Hermann Jacobi (1801–1874), the inventor of electrotyping. Once, when mistakenly taken for his brother, the mathematician Carl replied, "Pardon me, sir, I am not me. I am my brother."

The confusion between the Jacobi brothers was well known to many of Carl's friends, who, aware of Carl's penchant for droll humor, often jokingly took advantage of the mix-up. For example, Carl greatly admired the English astronomer and mathematician Mary Sommerville (1780–1872), and, when both were in Rome, he frequently called on her. Each time he did so, she would address him as, "Monsieur, votre frère [Sir, your brother]," which always caused Carl to chuckle.

Carl was a close friend of Professor and Mrs. Dirichlet. Once the latter gave Jacobi a Christmas package on which she had written, "al fratello del celebro Jacobi [to the brother of the celebrated Jacobi]."

On another occasion, when a very proper lady asked Carl whether he was a brother of the famous Jacobi, he replied, "Gracious, no, lady, he is my brother." Here one is reminded of the similar story, reported in Item 163° of *Mathematical Circles Squared,* about Henri Poincaré and his famous cousin Raymond Poincaré.

69° *Jacobi's drollery.*   Jacobi's drollery broke out on numerous occasions. Thus there was the time when he and a number of scholars attended a lecture given by Jacobi's friend, the astronomer F. W. Bessel (1784–1846). At the conclusion of the lecture, some of the scholars complained that Bessel had not really said anything new. Jacobi replied, "Yes, my friends, if Bessel had not explained his lecture to me during an hour-long walk this morning, I would not have understood it either."

—ELWOOD EDE

(freely translated from W. Ahrens, *Mathematiker-Anekdoten*)

70° *The Hävernick affair.*   Another instance of Jacobi's droll wit occurred when he was teaching at the University of Königsberg. At that time, the orthodox theologian Hävernick received an appointment to the "Albertina." Now the students disliked Hävernick because a decade earlier, while a student at Halle, he had served as an informant in the persecution of the rationalistic theologians Gesenius and Wegscheider. The result was that the students provoked an incident when Hävernick gave his first lecture at Königsberg. By plan, the entire student body crowded into the lecture hall, giving Hävernick a false feeling of great popularity, but as soon as Hävernick uttered his first syllable, the students stormed speedily and noisily out of the hall, leaving the deflated lecturer alone at his podium. The incensed Hävernick caused a hearing of the affair to be held by the University Senate, with an aim toward dismissal of the rude students. The prorector of the University, the anatomist Burdoch, concocted an apology to Hävernick, claiming that it was the steam and intense heat of the room that caused the students to rush out of the hall. At this point, Jacobi commented that, anyway, since only three students are needed to comprise a college, only the last three to leave the hall should be punished for dissolving the college.

—ELWOOD EDE

(freely translated from W. Ahrens, *Mathematiker-Anekdoten*)

71° *A Jacobi tactic.*   Jacobi had little respect for arrogant or pompous people. He once claimed, "If I have to speak with a tyrant or a foreign minister, I take pains to disappoint him beforehand."

72° *A gifted student.*   Jacobi was a brilliant student and won the admiration of all his teachers. The wide range of his talents is illustrated by the fact that in those early years he composed poetry. He once recited one of his own poems at a school assembly; the poem was written in Greek.

73° *Advice to a student.*   S. P. Thomson, in his *Life of Lord Kelvin* (1910), tells the following amusing story about the assistant to the famous mathematical-physicist Sir William Thomson (Lord Kelvin) (1824–1907).

The father of a new student, when bringing his son to the University, drew Professor Thomson's faithful assistant, Donald McFarlane, to one side and besought the assistant to tell him what his son must do to stand well with the professor. "You want your son to stand weel with the Profeessorr?" asked McFarlane. "Yes," replied the father. "Weel, then, he must have a guid bellyful o' mathematics!"

74° *Obvious, but for a different reason.*   Sir William Thomson (Lord Kelvin) often, in his writings, prefaced· some mathematical statement with the remark "it is obvious that" to the perplexity of his mathematical readers, for the statement was certainly not obvious from the mathematics that preceded it on the page. To Thomson the statement was obvious for *physical reasons,* and these reasons usually did not suggest themselves to the mathematicians, however competent they might be.

75° *A law of life?*   In his charming *A Budget of Paradoxes,* Augustus De Morgan (1806–1871) narrates the following whimsical anecdote.

"A few days afterwards, I went to him [the actuary referred to in Item 306° of *In Mathematical Circles*] and very gravely told him that I had discovered the law of human mortality in the Carlisle Table, of which he thought very highly. I told him that the law was involved in this circumstance. Take the table of the expectation of life, choose any age, take its expectation and make the nearest integer a new age, do the same with that, and so on; begin at what age you like, you are sure to end at the place where the age past is equal, or most nearly equal, to the expectation to come. 'You

37

don't mean that this always happens?'—'Try it.' He did try, again and again; and found it as I said. 'This is, indeed, a curious thing; this *is* a discovery!' I might have sent him about trumpeting the law of life: but I contented myself with informing him that the same thing would happen with any table whatsoever in which the first column goes up and the second goes down; . . ."

76° *De Morgan anagrams.*   De Morgan relates, in his *A Budget of Paradoxes,* that someone devised 800 anagrams on his name, of which he had seen about 650. Most of the anagrams are in Latin; here are two that are in English:

Great gun! do us a sum!
Go! great sum! $\int a^{u^n} du$

The first he regarded as a sneer at his pursuit and the second as more dignified.

77° *A problem in projectiles.*   There have been instances wherein a good mathematical problem has been suggested by some passage in a piece of literature. Perhaps the most famous such instance, historically, is the problem of the duplication of a cube, which seems to have been first suggested by a passage written by an unschooled and obscure Greek poet telling the story of the mythical King Minos and the cubical tomb erected for his son Glaucus. And one recalls how Euler was led to a problem in mechanics by a line in Virgil's *Aeneid.* These instances are told in Item 246° of *In Mathematical Circles,* along with that of an intriguing geometry problem that appeared some years ago in the Problem Department of *The American Mathematical Monthly,* suggested by some lines in Jan Struther's *Mrs. Miniver.* There are so many such instances that one could write an interesting little monograph on problems arising from literature. Here is another one, found among William Walton's *Collection of Mechanical Problems* and stemming from *The Song of Hiawatha* by Henry Wadsworth Longfellow (1807–1882). This popular poem was published in 1855. Based on Henry Rowe Schoolcraft's two informative books on the Indian tribes of North America and written in the catching trochaic metrics of the Finnish epic *Kalevala,* the poem was an immediate success. Here are the lines of the poem that caught William Walton's attention back in the mid-nineteenth century.

> Swift of foot was Hiawatha;
> He could shoot an arrow from him,
> And run forward with such fleetness,
> That the arrow fell behind him!
> Strong of arm was Hiawatha;
> He could shoot ten arrows upward,
> Shoot them with such strength and swiftness,
> That the tenth had left the bow-string
> Ere the first to earth had fallen.

From the above poetic lines, Walton proposed the following problem:

Suppose Hiawatha to have been able to shoot an arrow every second, and, when not shooting vertically, to have aimed so that the flight of the arrow might have the longest range. Prove that it would have been safe to bet long odds on him if he entered the Derby.

CHARLES HERMITE

THE two fundamental mathematical results due to Charles Hermite that are of most popular interest are his solution in 1858 of the general quintic equation by means of elliptic functions, and his proof in 1873 of the transcendence of the number *e.* Hermite's success with the quintic equation later led to the fact that a root of the general equation of degree *n* can be represented in terms of the coefficients by means of Fuchsian functions, and the method he employed to prove that *e* is transcendental was employed by Lindemann in 1882 to prove that π also is transcendental.

Charles Hermite was born in Dienze on December 24, 1822, and he died in Paris on January 14, 1901. Though not a prolific writer, most of his papers deal with questions of great importance

39

and his methods exhibit high originality and wide usefulness. His collected works, edited by Émile Picard, occupy four volumes. Some facts about him of a more personal nature can be learned from the following seven anecdotes, which may make up for the grave deficiency of his almost complete absence in our previous three trips around the mathematical circle.

78° *Hermite's infirmity.*   Mankind has always been plagued by infirmities, and the mathematicians have not been exempt. Recall that Saunderson was blind almost from birth, that Euler spent the latter part of his life in blindness, that Clifford and Abel died early of tuberculosis, that Hilbert suffered from pernicious anemia. The list can be extended.

Charles Hermite (1822–1901) was born with a deformity of his right leg and was accordingly lame all his life, requiring a cane to get about. An infirmity can easily mar one's disposition. Such, however, was not the case with Hermite, who uniformly maintained the sweetest of dispositions, causing him to be loved by all who knew him. One good result of Hermite's deformity was that it very successfully barred him from any kind of a military career. One bad result was that after one year at the École Polytechnique he was dropped from further study because the authorities claimed that his lame leg rendered him unfit for any of the positions open to successful students of the school.

79° *Hermite's kindness to those climbing the ladder.*   A number of eminent mathematicians have exhibited great generosity to younger men struggling for recognition. Charles Hermite is regarded as unquestionably the finest character of this sort in the entire history of mathematics.

80° *The yardstick of examinations.*   Hermite and Galois, both misfit alumni of the Louis-le-Grand lycée, exhibited a great indifference to the elementary mathematics of the classroom, and accordingly, consistently did poorly on all their school examinations. Professor Richard of the school had tried to save Galois, but failed; he also came to the rescue of Hermite, and this time fortunately succeeded. Anyone who detests examinations will feel for Hermite and Galois, and the careers of these two outstanding mathematicians may create some doubts in the minds of those who advocate examinations as a reliable yardstick for measuring one's intellectual merit.

While at Louis-le-Grand, Hermite had two papers, one of quite exceptional quality, accepted by the *Nouvelles annales de mathématiques,* a journal founded in 1842 and devoted to the interests of students in the higher schools. Professor Richard felt compelled to confide to Hermite's father that Charles was "a young Lagrange."

Later in 1842, while still twenty years of age, Hermite took the entrance examinations for the École Polytechnique, and although he was unquestionably the mathematical superior of some of his examiners, he had the humiliating experience of coming in sixty-eighth in order of merit, just barely meeting the entrance requirements. Ironically enough, Hermite's first academic post was an appointment, in 1848, as an examiner for admissions at the École Polytechnique, the very school that five years earlier had almost failed to admit him.

81° *A slight variation.* Many young mathematicians have married daughters of their professors. The reader may recall, from Item 322° of *Mathematical Circles Squared,* the resulting peculiar law of genetics: "Mathematical ability is not inherited from father to son, but from father-in-law to son-in-law." Hermite varied the usual procedure somewhat; he married, not a daughter, but the sister of one of his professors—in 1848 Hermite married Professor Bertrand's sister Louise.

82° *Hermite's conversion.* In the early part of his life Hermite, like many French scientists of his time, was a tolerant agnostic. But in 1856 he suddenly fell dangerously ill. While in his debilitated condition, Cauchy, who had deplored the young man's open-mindedness in religion, worked on the weakened Hermite and converted him to Roman Catholicism. Hermite lived the rest of his life a devout and contented Catholic.

83° *Hermite's mysticism.*   The whole question of mathematical existence is a highly controversial one. For example, do mathematical entities and their properties already exist in a sort of timeless twilight land of their own, and we, wandering about in that land, accidentally discover them? In this twilight land the medians of a triangle are, and always have been, concurrent in a point trisecting each median, and someone, probably in ancient times, wandering about in his mind in the twilight land, came upon this already existing property of the medians of a triangle. In the twilight land, many other remarkable properties of geometrical figures have always existed, but no one has yet stumbled upon them, and may not for years, if ever. In the twilight land, the natural numbers and their host of pretty properties already exist, and always have, but these properties will become existent in the real land of man only when someone wandering about in the twilight land comes upon them.

Pythagoras entertained the above idea of mathematical existence, as have many mathematicians after him. Hermite was a confirmed believer in the twilight land of mathematical existence. To him, numbers and all their beautiful properties have always had an existence of their own, and occasionally some mathematical Columbus stumbles upon one of these already existing properties and then announces his *discovery* to the world.

For more on discovery versus invention in mathematics, the reader may consult Item 355° of *Mathematical Circles Squared.*

84° *Hermite's internationalism.*   In this world of internecine strife and nationalistic jealousies, too many scientists look upon the scientific accomplishments of an enemy land as *bad* science, while that of their own land is *good* science. This narrow viewpoint has not bypassed the mathematicians. Thus, during World War II, many mathematicians of the allied nations looked upon "German" mathematics as somehow deformed and inferior, and many of the German mathematicians looked upon "Jewish" mathematics as somehow pernicious and evil. Hermite could never understand such politically partisan views of mathematics and science, and though he was a strong French patriot, to him mathematics and science were just mathematics and science, with no national or religious confines.

85° *On Cayley.* Writing in the *Comptes Rendus,* t. 120 (1895), p. 234, shortly after Cayley's death, Hermite said: "The mathematical talent of Cayley was characterized by clearness and extreme elegance of analytical form; it was reinforced by an incomparable capacity for work which has caused the distinguished scholar to be compared with Cauchy."

LEWIS CARROLL

A L L mathematicians admire Lewis Carroll (Charles Lutwidge Dodgson). Here are a few more stories about him.

86° *Lewis Carroll and the kitten.* Lewis Carroll (1832–1898) had a great empathy for animals. One time, when away from home, he came upon a kitten with a fishhook caught in its mouth. He carried the kitten to a doctor's house, where he held and comforted the animal while the doctor removed the hook by first snipping off the barbed end. Upon learning that the kitten did not belong to Carroll, the doctor declined payment. Lewis Carroll then carried the kitten back to the street where he had found it.

87° *Lewis Carroll and the horses.* On another occasion, Lewis Carroll came upon some horses that were being worked with checkreins on them. The discomfort to the horses caused by the checkreins was evident and instantly roused Carroll's compassion. He spoke to the man working the horses, and so convincingly put the case against the use of checkreins that the man removed them. The horses, allowed the natural use of their necks, performed their work much more efficiently and quickly.

88° *Lewis Carroll and meat.*   In Lewis Carroll's day, the butchering of animals for meat was done with little regard for the animals' physical feelings, and Carroll urged that methods of painless death for animals be adopted in the butchering process.

89° *An incongruous remedy.*   Professor York Powell, Regius Professor of Modern History at Oxford, has recorded a comic story told to him by Professor Dodgson (Lewis Carroll). A small child on being put to bed called to its nurse, "Nursey, my feet, my feet." The nurse took the child out of its cot and bathed his legs with vinegar and hot water, and gave him some warm milk. Upon putting the child back to bed, it once again cried out, "Nursey, my feet, my feet." So the nurse once again removed the child from his cot, but could find nothing amiss. To be on the safe side, however, she again bathed the child's legs with vinegar and hot water and dried them very carefully, and then put the now very sleepy youngster back into bed again. Immediately came the cry, "Nursey, my feet, my feet." So she took a light to examine the cot and discovered that the child's older brothers had, as a prank, short-sheeted the bed so that the child could not comfortably straighten out his legs.

90° *Lewis Carroll's fireplace.*   There was a fireplace in Lewis Carroll's study at Christ Church. Around the two sides and the top of the fireplace were sixteen square tiles and, across the middle of the top, a long rectangular tile. The square tiles were designed with pictures of animals, and the long tile above the fireplace-opening contained a ship. When young people visited Carroll in his study he would make the animals on the tiles carry on long and amusing conversations among themselves. He explained that the bird with its beak running through a fish, the dragon hissing defiance over its left shoulder, and others, represented the various ways in which he was accustomed to receive his guests.

The ship in the center above the fireplace became the famous vessel that the Bellman steered, often with difficulty, since "the bowsprit got mixed with the rudder sometimes." But

"The principal failing occurred in the sailing,
And the Bellman, perplexed and distressed,
Said he *had* hoped, at least, when the wind blew due East,
That the ship would *not* travel due West!"

The animals in the square tiles came to play a part in a number of Carroll's works. Thus in the bottom right corner was the Beaver, the only animal that the Butcher in *The Hunting of the Snark* knew how to kill. In the top right corner was the Eaglet, one of the competitors in the Caucus Race in *Alice in Wonderland,* and below it was the Gryphon. In the two uppermost tiles on the left side were the Lory and the Dodo, also of Caucus Race fame. The bottom left tile showed the Fawn that couldn't remember its name in *Alice Through the Looking Glass.*

QUADRANT TWO

From the antithesis of Plato
to Artin on graphs

A MELANGE

WE enter the second quadrant with stories concerning an assortment of men—from Francis Bacon to G. H. Hardy.

91° *The antithesis of Plato.* Thomas Macaulay, in his essay of 1837 on Francis Bacon (1561–1626), says: "Assuming the well-being of the human race to be the end of knowledge, he [Lord Bacon] pronounced that mathematical science could claim no higher rank than that of an appendage or an auxiliary to other sciences. Mathematical science, he says, is the handmaid of natural philosophy; she ought to demean herself as such; and he declares that he cannot conceive by what ill chance it has happened that she presumes to claim precedence over her mistress." And then, a bit further on, Macaulay supports Bacon by saying: "If Bacon erred here [in valuing mathematics only for its uses], we must acknowledge that we greatly prefer his error to the opposite error of Plato. We have no patience with a philosophy which, like those Roman matrons who swallowed abortives in order to preserve their shapes, takes pains to be barren for fear of being homely."

In truth, both Bacon and Macaulay were quite ignorant of the real nature and accomplishments of mathematics. As Augustus De Morgan said in his *A Budget of Paradoxes* of 1872, "If Newton had taken Bacon for his master, not he, but somebody else, would have been Newton."

92° *Hard to believe.* Most of our knowledge of Samuel Johnson (1709–1784), the great critic, man of letters, and lexicographer, has come to us from the remarkable biography of him written by his friend, the lawyer James Boswell. In this biography Boswell says, "When Dr. Johnson felt, or fancied he felt, his fancy disordered, his constant recurrence was to the study of arithmetic."

93° *A contrast of styles.* W. W. Rouse Ball, in his engaging *A Short Account of the History of Mathematics,* interestingly contrasts the styles of Lagrange, Laplace, and Gauss. "The great masters of modern analysis are Lagrange, Laplace, and Gauss, who were contemporaries. It is interesting to note the marked contrast in their styles. Lagrange is perfect both in form and matter, he is careful to explain his procedure, and though his arguments are general they are easy to follow. Laplace on the other hand explains nothing, is indifferent to style, and, if satisfied that his results are correct, is content to leave them either with no proof or with a faulty one. Gauss is as exact and elegant as Lagrange, but even more difficult to follow than Laplace, for he removes every trace of the analysis by which he reached his results, and studies to give a proof which while rigorous shall be as concise and synthetical as possible."

94° *Overshadowed.*   There have been occasions when an eminent mathematician was, during his lifetime, overshadowed by a much less gifted relative. Recall Henri Poincaré and his cousin Raymond Poincaré (see Item 162° of *Mathematical Circles Squared*) and C. G. J. Jacobi and his brother M. H. Jacobi (see Item 344° of *In Mathematical Circles* and Item 68° of the present volume). Another such instance is that of Augustus Ferdinand Möbius and his son. The father was born in Saxony in 1790 and, after an initial misstart preparing for the law, turned to mathematics. As a student of Gauss in 1813–14, he received training in astronomical observation and calculation. In 1815 he became a lecturer at the University of Leipzig, where he remained the rest of his life. In 1816 he became Extraordinary Professor of Astronomy and then, later, the Director of the Pleissenburg Observatory. In 1844 he was named Professor of Higher Mathematics and Astronomy at the university. He died, after a distinguished career, in 1868.

According to Heinrich Tietze, Möbius had a son who became a well-known neurologist, and whose book on the "physiologically weaker mind of women" won much more notice than did the sounder mathematical works of his father.

95° *A bright pupil.*   Even as a student at the Johanneum in Lüneberg, Riemann had so distinguished himself in mathematics that his mathematics instructor, Schmalfuss, allowed the boy to occupy himself in any manner he chose during the mathematics hour. Schmalfuss, an able mathematician in his own right, realized that Reimann had already mastered everything in mathematics that the school could offer.

96° *A perfectionist.*   Riemann greatly excelled in all of his subjects at the gymnasium, and he was indisputably very diligent, but, because he was a perfectionist, he rarely handed in any of his important assignments on time. His dormitory resident, a stern master, stood over the boy to be sure he didn't loaf. Still the assignments were late. Constantly dissatisfied with the quality of his work, Riemann would tear up one paper after another, and proceed anew. As a consequence, punishment was regularly meted out to the youngster, and he was often sent to the school prison cell to complete a late assignment. While thus incarcerated, he would ultimately produce a near-perfect paper, and he consistently earned top grades.

Riemann exhibited his trait of slow and patient striving for perfection in later years. He took an unusually long time to complete his doctorate to his own satisfaction, with the result that it was also a long time before he was promoted to a full professorship at Göttingen University.

97° *A beautiful result.*   Lazarus Fuchs (1833–1902), of differential equations fame, was born near Posen. He began his long teaching career at the University of Berlin, then taught at the Universities of Greifswald, Göttingen, and Heidelberg, and returned to Berlin as a full professor in 1884. It was in 1865 that he initiated a new theory of linear differential equations.

There is a frequently told story about a lecture given by Fuchs in his course on differential equations. It seems that he often appeared in class with his lecture unprepared, preferring to work out details at the blackboard. The result was that he occasionally made errors or cornered himself. During one of his lectures, while trying to derive a certain relation that he needed, he filled the board with a long series of complicated equations obtained by a sequence of involved substitutions. All at once the mass of equations simplified, and Fuchs suddenly arrived at the identity $0 = 0$. Puzzled and embarrassed, the professor paced a bit before the blackboard, muttering that there must be something wrong somewhere. Finally he turned to the class, winked, and said, "But no, zero equals zero is really a very beautiful result." Here the lecture terminated. At the following class meeting, Fuchs circumvented the whole problem by simply writing down the result he had originally wanted, and proceeded from there.

98° *A versatile genius.*   Hermann Ludwig Ferdinand von Helmholtz was born in Potsdam in 1821. He started his career as a student of medicine in Berlin and was appointed assistant surgeon in the Charité Hospital there in 1842. The following year he went to Potsdam as a military surgeon, but returned to Berlin in 1848 to teach anatomy at the Academy of Art. In 1849 he was called to the chair of physiology at Königsberg. Six years later he moved to Bonn as professor of anatomy and physiology. In 1858 he was appointed professor of physiology at Heidelberg. In physiology he did pioneering work on the human nervous system and was the first to measure the speed of nerve impulses. Among other things, he invented the ophthalmoscope used by doctors to examine eyes. His research work in physiology demanded so much physics that he applied himself to that subject and became an outstanding physicist. In physics he conducted pioneering research in acoustics and in optics, helped to prove the law of conservation of energy, and developed electromagnetic theory. In 1871 he returned to Berlin as professor of physics, and in 1888 he became president of the Imperial Physics-Technical Institute in Charlottenburg. His creative work in physics made such demand on mathematics that he became a high-ranking mathematician of his time. In addition to work in applied mathematics, Helmholtz contributed to the field of geometry, particularly to non-Euclidean geometry. He died in Charlottenburg in 1894.

Of Helmholtz, William Kingdon Clifford wrote: "Helmholtz—the physiologist who learned physics for the sake of his physiology, and mathematics for the sake of his physics, and is now in the first rank of all three."

99° *Groping.* Alfred North Whitehead (1861–1947) once cautioned a student about a theory of logic. "You must take it with a grain of, er . . . um . . . ah . . ." For almost a minute Whitehead groped for the word, until the student suggested, "Salt, Professor?" "Ah, yes," Whitehead beamed, "I knew it was some chemical."—LEO MOSER

100° *Scepticism.* It has been claimed that only three people read the three volumes of Russell and Whitehead's monumental *Principia mathematica* in its entirety, namely Russell, Whitehead, and the proofreader. There has been some scepticism, however, about Russell and Whitehead belonging to the list.

—LEO MOSER

101° *Hilbert's memory.* Helmut Hasse, when on a visit to the University of Maine in March, 1972, told an amusing story about himself and David Hilbert (1862–1943). According to the story, Hasse once expressed to Mrs. Hilbert a desire to speak personally with the great mathematician. Mrs. Hilbert accordingly invited Hasse to tea one afternoon and then left him in the garden with her husband. Hasse soon launched into a discussion of class-field theory, a subject that had been created by Hilbert and that was of great interest to Hasse at the time. Hasse had written a report in the theory that had continued the earlier work done by Hilbert, and Hasse started to tell Hilbert of his added contributions. But Hilbert repeatedly interrupted Hasse, insisting that Hasse first explain the basic concepts and foundations of class-field theory. Hasse did so, and Hilbert grew enthusiastic and finally exclaimed, "All this is extremely beautiful; who created it?" And Hasse had to tell the astonished Hilbert that it was Hilbert himself who had created the beautiful theory.

One is reminded of the similar story told of J. J. Sylvester (see Item 353° of *In Mathematical Circles*).

102° *Hilbert and his boiled eggs.* During World War I, Hilbert was met on his way to the University by a colleague. Asked by the colleague, "How are you, Professor Hilbert?" he replied, "Oh, rather bad." And when further asked, "What is the matter with you?" he said, "My wife is trying to murder me." The colleague was astonished to hear this. He said he could not believe it, because he knew how anxious Mrs. Hilbert was about her husband's welfare. But Hilbert told him that since their marriage his wife had boiled him *two* eggs each morning for breakfast, and the colleague would certainly understand how shocked he was this morning to get only *one* egg. (It was at the time when food rationing had set in in Germany.)—HELMUT HASSE

103° *Hilbert and his torn trousers.*   Hilbert used to go on a bicycle excursion, on a Saturday afternoon, with the students of his Seminar during the strawberry season. Now it was about this time of the year that his students had seen the Professor day by day coming to the University with a large rent in the seat of his trousers. The great respect they had for the Professor prevented them from drawing his attention to the tear. So they decided to wait until the impending excursion to the nearby village of Nikolausberg, where they were to revel in fresh strawberries and cream, and they then would say to him, as he descended from his bicycle, "Herr Professor, you have just now, in getting off your bicycle, torn your trousers." They did as planned, but Hilbert immediately answered, "But no, gentlemen, I have had this already for a fortnight."—HELMUT HASSE

104° *An unusual visit.*   Jacques Hadamard (1865–1963) has told that, as editor of a mathematics journal, he once received rather good papers from someone unknown to him, so he invited the person to dinner. The correspondent wrote that owing to circumstances beyond his control he could not accept the invitation, but he invited Hadamard to visit him. Hadamard did so and found to his great surprise that the author was confined to a criminal lunatic asylum. Apparently he was quite sane except for the murder of his aunts. His name was A. Bloch, and he was a very good mathematician.—L. J. MORDELL

105° *A conscientious reading.*   Edmund Landau (1877–1938) was once sent a thesis to referee. He did not return it for a long time and several months later the sender met Landau and asked him if he had read the work. Landau replied, "Ya, Ich habe es grundlich durchgebletert [Yes, I have thoroughly paged through it]."—LEO MOSER

106° *Place cards.*   Landau gave an annual banquet for his graduate students. Instead of place cards he made up cards containing some characteristic phrase or formula from the subject of the student's thesis. One student had made no progress whatever towards a thesis topic and he wondered what would appear on his card. It turned out that he had no difficulty in finding his place. One card was quite blank.—LEO MOSER

107° *Continuity.*   Professor G. H. Hardy (1877–1947) did not believe in wasting time reviewing his earlier lectures—the students were expected to retain in their minds the material already covered. It chanced that a series of mathematical lectures given by Professor Hardy at Cambridge University was interrupted by the long vacation. On the first class meeting after the vacation, Professor Hardy advanced to the blackboard with chalk in hand and said, "It thus follows that. . . ."—STANLEY B. JACKSON

108° *Hundred percent achievers.*   When I visited Hardy in his rooms in Trinity College in 1933, I saw three pictures on his mantelpiece, namely, one of Lenin, one of Jesus Christ, and one of Hobbs (the then star cricketer in England). Asked why there were exactly those three, Hardy said they were, in his opinion, the only important personalities who had achieved a hundred percent of what they wanted to achieve.—HELMUT HASSE

109° *Hardy tries to outsmart God.*   [In Item 298° of *Mathematical Circles Revisited,* we told George Polya's version of a story of how Hardy once attempted to outsmart God. Good stories often appear with variations, and this is true of the present story. Here is Leo Moser's version of the story; it replaces the Riemann hypothesis of the former telling by Fermat's last theorem.]

Hardy claimed that he believed God existed but that God was his personal enemy. Once Hardy had to travel by air to France to attend a meeting. He feared air travel and so was very reluctant to go. Finally he did so, but he left a note on his desk to be opened by his secretary after his departure. When the note was opened, it was found to read: "I have just discovered a proof of Fermat's last theorem and will supply the details upon my return from France." His colleagues were quite excited about this and when Hardy came back they immediately pounced upon him. "What about the proof?" they asked. "Oh, that!" Hardy replied, "That was just insurance. I felt sure that God would spare me on this trip in order that no undue credit would go to me for having solved Fermat's last theorem."—LEO MOSER

ALBERT EINSTEIN

PROBABLY no mathematician of modern times has been more universally known and admired than Albert Einstein (1879–1955), and stories about this great man are always welcomed. Here are some more to add to those we have already told in our earlier trips around the mathematical circle.

110° *Home-town honors.* Albert Einstein was born in Ulm, Germany, in 1879. Though the Einstein family moved from Ulm to Munich when Albert was only about a year old, one nevertheless wonders if the city of birth of the famous mathematical physicist has in any way honored its most outstanding citizen. For example, one might expect that the house in which Einstein was born has been preserved as an historical shrine, but that house was blasted to rubble in World War II and thus no longer stands. However, a street in Ulm was named Einsteinstrasse in honor of Einstein. But the Nazis could not bear to see a Jew honored in this way, and the new Nazi mayor of Ulm on his first day in office changed the name of the street to Fichtestrasse, after the eighteenth-century German philosopher, orator, and patriot Johann Gottlieb Fichte (1762–1814). Succeeding the defeat of the Nazis, the name Einstein-strasse was restored, and so it stands today.

111° *No sales resistance.*   There is a story about Einstein that says one day a large truck appeared before his modest two-story frame house in Princeton, New Jersey, containing a sizable elevator to be installed in the house. It seems that Einstein, who had absolutely no need or desire for such an expensive luxury, could never be rude to a salesman, and so some weeks earlier had allowed himself to be seduced into ordering the thing. Mrs. Einstein refused acceptance of the elevator and sent the loaded truck away. —ROBERT T. HALL

112° *The great stimulator.*   An enormous emotion was aroused in Albert Einstein very early in his twelfth year when he came into possession of a small textbook on Euclidean geometry. The book utterly absorbed his interests, and later, in his *Autobiographical Notes,* he wrote with rapture of "the holy little geometry book." He says: "Here were assertions, as for example the intersection of the three altitudes of the triangle in one point, which—though by no means evident—could nevertheless be proved with such certainty that any doubt appeared to be out of the question. This lucidity and certainty made an indescribable impression on me." We have here another instance of an eminent mathematician whose first great stimulus for mathematics was received from the enchanting material of Euclid's geometry.

113° *Mystery.*   One night in New York, scientist Albert Einstein attended a banquet given in his honor. Mrs. Einstein, ailing with a cold, did not accompany him.

It was a formal affair, with the men in white ties and the ladies in décolleté evening gowns.

When Einstein came home, he found his wife waiting up for him, eager to learn what had taken place. He began to tell her about the famous scientists who were present, but she cut him short.

"Never mind that," she said. "How were the ladies dressed?"

"I really don't know," replied Einstein. "Above the table, they had nothing on, and under the table I didn't dare look!"

—*Famous Fables* by EDGAR

114° *The weapons of World War IV.*   Someone once asked Einstein what weapons will be used in World War III. "I don't know," he replied, "but I do know what weapons will be used in World War IV." "What will those weapons be?" he was asked. "Sticks and stones," he replied.

115° *A considerate man.*   Dr. John Wheeler of Seattle, Washington, and one of America's leading theoretical physicists, taught at Princeton University during the early 1950s when Albert Einstein was also there. He recalls, "Dr. Einstein used to walk past my house on his way home from the university, and from time to time my children's cat would follow him home. As soon as he got there, he would phone to tell me that our cat was over at his house. He didn't want us to worry."

116° *A late bloomer.*   Albert Einstein was unable—or perhaps unwilling—to speak until he was three years old. In school he had no facility at sums and had to be taught the multiplication table by raps on the knuckles. He was so mediocre as a pupil that at seven his teacher said "nothing good" would ever come of him. When he sought admission to the Polytechnic Institute in Zurich (now the Federal Institute of Technology), he failed to pass the entrance examination and had to prepare for a second attempt at entrance. After graduating from the Institute, he managed to secure a modest position at the patent office in Berne; his efforts to obtain the equivalent of a high school teaching post were consistently turned down.

Einstein's ineptness in simple arithmetic followed him through life, but it is believed that his late use of verbal communication led him to develop an extraordinary capacity for nonverbal conceptualization, so that the use of abstract concepts, rather than words, persisted into his adult life. Dr. Gerald Holton, professor of physics at Harvard University, believed Einstein's habit, from infancy on, of thinking in concepts rather than words played a key role in Einstein's scientific work. Once, when discussing the genesis of his ideas with a friend, Einstein commented: "These thoughts did not come in any verbal formulation. I rarely think in words at all. A thought comes, and I may try to express it in words afterward." Elsewhere he wrote: "The words or the language, as they are written or spoken, do not play any role in my mechanics of thought." The elements of Einstein's thought seem to have been sets of images that he could voluntarily reproduce or combine, or, sometimes, certain muscular reminders would be used instead of visual symbols.

117° *Two mementos.* There were two objects, Einstein claimed in later years, that played a special inspiration in his life. One of these was the little geometry book mentioned above in Item 112°; the other was a directional compass given to him by his father when he was four or five years old. The little book enthralled him as a boy because of the beautiful certainty of the deductive reasoning revealed in it; the compass mystified him as a youngster because of the all-pervading magnetic field that cleverly controlled the needle. Einstein was so impressed by these two objects that he kept them close to him all his life.

What has become of these two objects that so influenced Einstein's life? The little book is still in existence—in the filing cabinets at Princeton; the compass, on the other hand, has somehow vanished.

118° *Declining a presidency.* Einstein exerted long efforts on behalf of the creation of a Jewish national state. On May 15, 1948, the dream came true with the establishment of the independent State of Israel in Palestine. In 1952 the Israeli government asked Einstein to accept the presidency of that country as successor to Chaim Weizmann. Einstein sadly declined the honor, insisting that he was not fitted for such a position.

59

119° *Einstein, the sage.*   In his later years, Einstein became fixed in the public eye as something of a philosophical sage, and many people wrote to him for advice concerning their personal problems. One of the most poignant of these letters was one that he received in 1950 from an ordained rabbi. The rabbi wrote that he had tried in vain to comfort his nineteen-year-old daughter over the death of her sixteen-year-old sister. The surviving daughter found no consolation "based on traditional religious grounds," and so the concerned father thought that perhaps a scientist could help.

"A human being," wrote Einstein in his reply, "is a part of the whole, called by us 'Universe,' a part limited in time and space. He experiences himself, his thoughts and feelings as something separated from the rest—a kind of optical delusion of his consciousness. This delusion is a kind of prison for us, restricting us to our personal desires and to affection for a few persons nearest to us. Our task must be to free ourselves from this prison by widening our circle of compassion to embrace all living creatures and the whole of nature in its beauty. Nobody is able to achieve this completely, but the striving for such achievement is in itself a part of the liberation and a foundation for inner security."

120° *The deleted passage.*   Einstein was an impassioned humanitarian and internationalist and he regarded nationalism and patriotism as generators of evil aggressiveness.

On October 23, 1915, at the height of World War I and while living in Berlin, Einstein was invited by the Berlin Goethe Society to submit an essay for publication in the Society's journal. The ensuing correspondence brought Einstein face to face with German nationalism, and finally forced him to delete a passage from his article. In reply to the invitation Einstein wrote, "Of course, I will not be surprised, or even indignant, if you do not make use of my remarks. However, in that case, I ask you to send the same back to me."

The Goethe Society was indeed dismayed by Einstein's submitted essay, for it contained a passage in which Einstein equated patriotism with the worst of aggressive animal instincts. The outcome was that he was asked to delete this part from his article, for the concept of patriotism was the chief prop then upholding the morale of the German troops bogged down in muddy trenches along the various fronts of the war. Here is the passage that Einstein finally agreed to delete:

"One may ask oneself why it is that in peacetime, when the social system suppresses almost all expressions of virile pugnaciousness, the attributes and drives that during war generate mass murder do not disappear. In that respect it seems to me as follows:

"When I look into the home of a good, normal citizen I see a softly lighted room. In one corner stands a well-cared-for shrine, of which the man of the house is very proud and to which the attention of every visitor is drawn in a loud voice. On it, in large letters, the word 'Patriotism' is inscribed.

"However, opening this shrine is normally forbidden. Yes, even the man of the house knows hardly, or not at all, that this shrine holds the moral requisites of animal hatred and mass murder that, in case of war, he obediently takes out for his service.

"This shrine, dear reader, you will not find in my room, and I would rejoice if you came to the viewpoint that in that corner of your room a piano or a small bookcase would be more appropriate than such a piece of furniture which you find tolerable because, from your youth, you have become used to it."

Continuing, Einstein wrote: "It is beyond me to keep secret my international orientation," and he concluded by saying that the state, to which he happens to belong as a citizen, "does not play the least role in my spiritual life; I regard allegiance to a government as a business matter, somewhat like the relationship with a life insurance company."

Thus ended the offending passage that Einstein finally agreed to delete from his essay. He concluded the article, however, by saying: "But why so many words, when I can say everything in one sentence, and also a sentence that suits my being a Jew: Honor your master, Jesus Christ, not only with words and hymns, but above all thoroughly by your deeds."

121° *First brush with patriotism.* Einstein said he could never forget the open hatred his classmates in grade school had, under a misguided sense of neighborhood patriotism, for the students of a nearby school. Numerous unreasonable fist fights took place, many heads were battered in the frenzied battles, and bloodied "heroes" were idolized.

122° *Einstein as the hero of an opera.* The prominent East German composer Paul Dessau for almost twenty years carried around the idea of composing an opera built about the figure of Albert Einstein. The opera was not to be biographical, nor a setting in music of relativity theory, but a sort of morality play on a scientist's responsibility for destructive discoveries that he lets loose upon the world. Finally, in 1974, when eighty years old, Dessau saw his opera completed and performed. Under superb direction and with excellent singers, the opera *Einstein* enjoyed a very successful world premier at the East Berlin Staatsoper, conducted by Otmar Suitner of the San Francisco Opera Company.

The opera is unconventional, but contains fascinating music and a comic allegoric episode provided by *Hanswurst* (Germany's *Punch*) after each act. A single setting cleverly used with simple props and scenic units evokes the various atmospheres of the rapidly changed nineteen scenes.

The opera opens with an alarming prologue informing the audience that the Nazis (all appearing with their arms bloody to the elbows) are burning undesirable books on the square just outside the Staatsoper (where the burning actually took place), a site "you can visit during the intermission," and Scene One shows the infamous incident. Einstein's apartment is vandalized by storm troopers and the mathematical physicist leaves for the United States with his violin under his arm. One of Einstein's colleagues follows him, and then returns to Germany as an American officer to impress the services of a second mathematical physicist, a Nazi collaborator and organ virtuoso, to develop the atomic bomb. Act II ends when the bomb is dropped on Hiroshima. In Act III Einstein and his fellow scientists, guarded by MPs carrying machine guns (all United States authorities, including "The President"—presumably Roosevelt—appearing with their legs bloody to the knees) are haled before the United States Supreme Court to answer for their conduct in condemning the bomb, and Einstein is "sentenced" to everlasting fame. Einstein's former colleague leaves to join the Communist camp, and Einstein, now a bitter old man, destroys his latest theories before they can be put to inhuman use. It is in the allegoric *Hanswurst* intermezzi closing each act that the moral of the opera is provided.

L. J. MORDELL

THERE are few things more entertaining to fellow mathematicians than when a famous figure in the field decides, after years of life and research, to reminisce in their presence, and to tell stories about himself and his work. L. J. Mordell, the renowned number theorist and expert in Diphantine analysis, has done this on a number of occasions. The following stories, along with others credited to him in this book, have been culled, with his gracious permission, from his article: "Reminiscences of an octogenarian mathematician," *The American Mathematical Monthly,* Nov. 1971. Though Mordell spent most of his professional life in England, he was born in America—in Philadelphia, in 1888—and lived his first eighteen years there.

123° *A couple of confessions.*   The great number theorist, L. J. Mordell, referring to his grammar-school days, said: "I think I was good at arithmetic, but certainly not in later life."

Recalling a mathematics test that he took in 1904, when he was sixteen years old, Mordell said: "All I remember about the examination is that there was a question on Sturm's theorem about equations, which I could not do then and cannot do now."

124° *The beginning of his major interest.*   About the age of fourteen, before entering high school, I came across some old algebra books in the five-and-ten-cent counters of Leary's famous bookstore in Philadelphia, and for some strange reason the subject appealed to me. One of these books was *A Treatise on Algebra* by C. W. Hackley, who was a professor from 1843 to 1861 at what was then called Columbia College in New York City. My copy was the third edition, dated 1849 (the first appeared in 1846). It was really a good book, though not rigorous, and contained a great deal of material, including the theory of equations, series, and a chapter on the theory of numbers. Like the old algebra books of those days, it had a chapter on Diophantine analysis, a subject I found most attractive. It is not without interest that in later years much of my best research deals with this subject. In fact, I have recently written a book on the subject,* which appeared in 1969.

—L. J. MORDELL

125° *History repeats itself.* In 1912, an international congress was held in Cambridge, and I attended it. I am fully aware of the implications of the story I am going to tell. I went into the buffet room where all the distinguished mathematicians were gathered, and I thought to myself, "What an odd-looking lot they are." I have no doubt that history repeats itself.—L. J. MORDELL

126° *On a train.* During the Second World War we had a country cottage at Chinley, about twenty miles from Manchester, where I was then a professor. My wife and I were coming home one weekend by train. We entered a compartment, and my wife sat diagonally opposite from me. In front of me was a youth and beside me a middle-aged man. Presently I noticed that the youth was reading a book entitled *Teach Yourself Trigonometry*. Hello, I thought, we are in the same profession. So I asked him whether it was an interesting subject. He did not reply, maintaining a stony silence. Obviously this was an important war secret, and Hitler was not going to get any information from him. Five minutes later I tried again and asked him whether it was a difficult subject. Again no reply, and so I tried no further. When we came to our local station, I got out, and my wife continued into town. She told me

---

*Diophantine Equations. New York: Academic Press, 1969.

afterwards what took place. The other man turned to the youth and said, "You were very rude. Why did you not answer the gentleman?" The reply was, "What does he think he knows about mathematics?"—L. J. Mordell

127° *Poor sales resistance.*   In 1953, when I was a Visiting Professor at the University of Toronto, I went with my wife to buy a pair of socks. By the time I left the store, she and the salesman persuaded me to buy an overcoat. When I related this to a doctor friend, he said he knew a far better salesman. A woman, whose husband had died, went to buy a suit of clothes to bury him in. The salesman persuaded her to buy a suit with two pairs of trousers. —L. J. Mordell

128° *The perfect reply.*   In 1958 I was a Visiting Professor at the University of Colorado at Boulder. One day the phone rang, and a woman's voice said, "I am Ann Lee and I want to give you a chance of winning forty-five dollars." I said, "Oh." She then asked me, "What is the oldest dance in the world?" I said, "That's a difficult question and I don't know the answer." She then asked me, "Where does the tango come from?" I replied, "South America." "Good," she said, "you have answered the question, and you are now entitled to forty-five dollars of free dancing lessons." You may think for a moment as to what reply you should make to this, but I would get top marks. I asked her, "Who was the first President of the United States of America?" "George Washington," she said. "Good," I replied, "you have won your forty-five dollars back again."—L. J. Mordell

129° *Recalling distinguished classmates.*   One of my classmates there [the Grammar School Modell attended in Philadelphia], now Professor F. C. Dietz, is Professor Emeritus of History at the University of Illinois at Urbana; I saw him in December last when I lectured there. I mentioned this to Professor P. Bateman, who is head of the Mathematical Department there, and he countered by saying that one of *his* classmates had been electrocuted for murder.—L. J. Mordell

130° *Leo Moser's toast to L. J. Mordell.*

> Here's a toast to L. J. Mordell,
> Young in spirit, most active as well.
> He'll never grow weary
> Of his love, number theory.
> The results he obtains are just swell.

FROM OUR OWN TIMES

THE following anecdotes are about mathematicians of our own times. Since most of these people are still living, it is with some trepidation that the stories are inserted, for it certainly is not our intention to injure the feelings of any of our contemporary brethren. Indeed, the inclusion of a story about a living or a recently deceased person, even though the story pokes fun at the person, should be interpreted as a sign of affection for that person. Most of these stories are from a collection begun several years ago by Leo Moser, who forwarded them to Leon Bankoff. After the sad and sudden death of Moser, Bankoff passed the anecdotes on to the present writer, who now offers them, along with a few others, for general perusal. A collection of stories about contemporary mathematicians can easily be expanded to fill a book of its own.

131° *A modest man.* In 1923 I attended a meeting of the American Mathematical Society held at Vassar College in New York State. Someone called Rainich,* from the University of Michigan at Ann Arbor, gave a talk upon the class number of quadratic fields, a subject in which I was then very much interested. I noticed that he made no reference to a rather pretty paper written by one Rabinowitz from Odessa and published in *Crelle's Journal.* I commented upon this. He blushed and stammered and said, "I am

---

*G. Y. Rainich, Emeritus Professor of Mathematics, University of Michigan.

Rabinowitz." He had moved to the United States and changed his name.—L. J. MORDELL

132° *Out of this world.*   One day, while lecturing, Professor Halmos made a mistake at the blackboard. He erased it, saying: "Excuse me. I am always in Hilbert space."—PETER GEDDES

133° *Please accommodate us.*   Of Leo Moser it was said that he was writing a book and taking so long about it that his publishers became very much worried and went to see him. He said he was very sorry about the delay, but he was afraid that the book might have to be a posthumous one. Well, he was told, please hurry up with it.—L. J. MORDELL

134° *Continued fractions.*   Helmut Hasse spoke yesterday on continued fractions. But, of course, he didn't finish.
                                        —CLAYTON W. DODGE

135° *Dan Christie.*   When Professor Dan Edwin Christie of Bowdoin College suddenly passed away on July 18, 1975, some of his close colleagues delivered short talks at the memorial services held four days later in the Bowdoin College Chapel.
   Professor James E. Ward III, in his talk, recalled the following little incident. In the preceding summer, Ward taught a course from the manuscript of Dan's last book, and dropped Dan a short note pointing out a minor error. Ward concluded his note by saying, "It's elegant, but it's not correct." Within a matter of hours, Ward received a note back from Dan acknowledging the error, offering an improvement to a correction suggested by Ward, and concluding with, "If it's incorrect, then it's not elegant."
   Professor Richard L. Chittim, who presented the final short talk, illustrated Dan's famous one-line retorts by a story reported by Professor Fritz Koelln. Once when Koelln and Christie were talking about childhood homes, and in particular about Milo, Maine, where Dan was born, Koelln laughingly asked Dan if he had known the Venus of Milo. Without hesitation Dan replied, "I went to school with her."

136° *Under the wheels.*   At the Physics Congress held in Zurich in 1931, P. L. Kapitza had himself photographed lying on the ground close to the wheels of a car. He explained, "I just want to know what I should look like if I were being run over."

—LEO MOSER

137° *Pride.*   While still undergraduates, the Austrian Houtermans and the Englishman Atkinson, during a walking tour near Göttingen, began to work out their theory of the thermonuclear reactions on the sun, a theory that later achieved much fame. The theory for the first time put forward the conjecture that solar energy might be attributed not to demolition but to fusion of lightweight atoms. The development of this theory led straight to the hydrogen bomb. At the time (1927), of course, neither of the two young students of the atom dreamed of such sinister consequences.

Houtermans reports: "That evening, after we had finished our essay, I went for a walk with a pretty girl. As soon as it grew dark, the stars came out, one after another, in all their splendor. 'Don't they sparkle beautifully?' cried my companion. But I simply stuck out my chest and said proudly: 'I've known since yesterday why it is that they sparkle.' "—LEO MOSER

138° *Erdös's new suit.*   Paul Erdös came to the University of Syracuse one day in a new suit. His colleagues were surprised. They were even more surprised when Erdös removed his jacket and they noticed that his vest had a rectangular hole cut out of it. "What is the meaning of this?" they inquired. Erdös explained, "There was a ticket sewn to the vest. The ticket just had the size and price of the suit on it and I didn't need these so I cut it out."
—LEO MOSER

139° *Naïveté.*   After seeing the movie *Cyrano de Bergerac* with some friends, Hans Zassenhaus commented, "An excellent picture. The producers were fortunate to find such an excellent actor who at the same time has such a long nose."—LEO MOSER

140° *A good reason.*   One cold and stormy day, J. Lambek was on his way to McGill University by taxi. He noticed Zassenhaus on a street corner waiting for a bus, so he had the taxi driver stop and he offered Zassenhaus a lift. "No, thank you," said Zassenhaus. "If I can't get to work on time by bus, then I don't want to get there on time at all."—LEO MOSER

141° *A reasonable excuse.*   A. Schild was a graduate student at the University of Toronto. He was working on a thesis on cosmology and, at the same time, attending, or at least supposed to be attending, various lecture courses, including one on topology. He missed many consecutive lectures in this latter course and finally, when he did show up for a lecture, the instructor asked him why he had been absent so long. His answer: "I have been busy calculating the size of the universe."—LEO MOSER

142° *An explanation.*   After Loo King Hua returned to Red China, there were (false) reports in some American newspapers that he had committed suicide. R. Ayoub, who was a student of Hua's, explained these reports as follows: "I sent him my Ph.D. thesis and I guess he just couldn't take it."
—LEO MOSER

143° *Progress.*   In 1951, E. Teller was asked: "Will the thermonuclear device work?" He replied, "I don't know."
"But if you didn't know five years ago, haven't you made any progress since then?"
"Oh, yes," Teller replied, "Now I don't know on much better grounds."—LEO MOSER

144° *Some new-old results.*   J. L. Synge once told A. Weinstein about some new results he had just obtained. Weinstein told Synge that the results were not new and indeed appeared in a recent issue of an Italian journal. Synge said, "That is impossible. I subscribe to this journal and I saw no such article." Weinstein stuck to his opinion, so they went to Synge's office and Synge pulled out the journal in question. Several pages were joined together and when these were separated the paper that Weinstein had mentioned emerged.—LEO MOSER

145° *Poetry and physics.* One evening Paul Dirac, who was usually so silent, took Oppenheimer aside and gently reproached him. "I hear," he said, "that you write poetry as well as working in physics. How on earth can you do two such things at once? In science one tries to tell people, in such a way as to be understood by everyone, something that no one ever knew before. But in poetry, it's the exact opposite!"—LEO MOSER

146° *Amazing.* If two integers are chosen at random, the probability that they are relatively prime is $6/\pi^2$. In an article in the *Scientific American,* July 1953, pp. 31–35, entitled "Circumetrics," N. T. Bridgeman describes an experimental verification of this fact as follows:

"Thus Chartres, in 1904, made a random selection of 250 pairs of primes and found that 154 of them were prime to each other; and as it is known that the probability of the conjunction is $6/\pi^2$, his trial amounted to an estimate of π of 3.12."

—LEO MOSER

147° *A noble ambition, but....* Howard Fehr, on being told by H. S. M. Coxeter that one of Coxeter's children was considering studying for the ministry, remarked, "A noble ambition, but I trust he will grow out of it."—LEO MOSER

148° *Yes, but.* Leopold Infeld, of the University of Toronto, gave a course in the philosophy of science. At one time he was discussing new concepts introduced into physics in connection with various theories of the origin of the universe. A student asked, "Can one not similarly postulate the existence of God?"

"Of course, of course," said Infeld, "but, my boy, you know quite well that this is not the sensible thing to do."

—LEO MOSER

ON MATHEMATICS AND MATHEMATICIANS

149° *On mathematicians.*   The good Christians should beware of mathematicians and all those who make empty prophecies. The danger already exists that the mathematicians have made a covenant with the Devil to darken the spirit and to confine man in the bonds of Hell.—ST. AUGUSTINE

150° *An element of luck.*   There is an old adage: Oats and beans and barley grow, but neither you nor I nor anybody else knows what makes oats and beans and barley grow. Neither you nor I nor anybody else knows what makes a mathematician tick. It is not a question of cleverness. I know many mathematicians who are far abler and cleverer than I am, but they have not been so lucky. An illustration may be given by considering two miners. One may be an expert geologist, but he does not find the golden nuggets that the ignorant miner does.

—L. J. MORDELL

151° *Existence proofs.*   God exists since mathematics is consistent, and the Devil exists since we cannot prove it.

—A. WEIL

152° *The game of mathematics.*   God is a child; and when he began to play, he cultivated mathematics. It is the most godly of man's games.

—V. ERATH
Das blinde Spiel (1954)

153° *An analogy.*   A student who has merely done mathematical exercises but has never solved a mathematical problem may be likened to a person who has learned the moves of the chess pieces but has never played a game of chess. The real thing in mathematics is to play the game.—STEPHEN J. TURNER

154° *Longevity.* Archimedes will be remembered when Aeschylus is forgotten, because languages die and mathematical ideas do not.

—G. H. HARDY
A Mathematician's Apology

155° *The inner self.* In some ways, a mathematician is not responsible for his activities. One sometimes feels there is an inner self occasionally communicating with the outer man. This view is supported by statements made by H. Poincaré and J. Hadamard about their researches.* I remember once walking down St. Andrews Street some three weeks after writing a paper. Though I had never given the matter any thought since then, it suddenly occurred to me that a point in my proof needed looking into. I am very grateful to my inner self for his valuable help in the solution of some important and difficult problems that I could not have done otherwise.—L. J. MORDELL

156° *A felicitous motto.* Ever since issue No. 26, 1901, *The Mathematical Gazette* (the official journal of the Mathematical Association, an association of teachers and students of elementary mathematics in Great Britain) has carried on its cover the apt motto of Roger Bacon:

> I hold every man a debtor to his profession, from the which as men of course do seek to receive countenance and profit, so ought they of duty to endeavor themselves by way of amends to be a help and an ornament thereunto.

157° *Mathematics and Antaeus.* The mighty Antaeus was the giant son of Neptune (god of the sea) and Ge (goddess of the earth), and his strength was invincible so long as he remained in contact with his mother earth. Strangers who came to his country were forced to wrestle to the death with him, and so it chanced one day that Hercules and Antaeus came to grips with one another. But

---

*See Items 348°, 349°, 350°, 351°, 353°, and 358° in *Mathematical Circles Squared.*

Hercules, aware of the source of Antaeus' great strength, lifted and held the giant from the earth and crushed him in the air.

Surely there is a parable here for mathematicians. For just as Antaeus was born of and nurtured by his mother earth, history has shown us that all significant and lasting mathematics is born of and nurtured by the real world. As in the case of Antaeus, so long as mathematics maintains its contact with the real world, it will remain powerful. But should it be lifted too long from the solid ground of its birth into the filmy air of pure abstraction, it runs the risk of weakening. It must of necessity return, at least occasionally, to the real world for renewed strength.

158° *The tree of mathematics.* [Adapted from Section 15–10 of Howard Eves, *An Introduction to the History of Mathematics,* fourth edition. New York: Holt, Rinehart and Winston, 1976.]

It became popular some years ago to picture mathematics in the form of a tree, usually a great oak tree. The roots of the tree were labeled with such titles as *algebra, plane geometry, trigonometry, analytic geometry,* and *irrational numbers.* From these roots rose the powerful trunk of the tree, on which was printed *calculus.* Then, from the top of the trunk, numerous branches issued and subdivided into smaller branches. These branches bore such titles as *complex variables, real variables, calculus of variations, probability,* and so on, through the various "branches" of higher mathematics.

The purpose of the tree of mathematics was to point out to the student not only how mathematics had historically grown, but also the trail the student should follow in pursuing a study of the subject. Thus, in the schools and perhaps the freshman year at college, he should occupy himself with the fundamental subjects forming the roots of the tree. Then, early in his college career, he should, through a specially heavy program, thoroughly master the calculus. After this is accomplished, the student can then ascend those advanced branches of the subject that he may wish to pursue.

The pedagogical principle advocated by the tree of mathematics is probably a sound one, for it is based on the famous law pithily stated by biologists in the form: "Ontogeny recapitulates phylogeny," which simply means that, in general, "The individual repeats the development of the group." That is, at least in rough outline, a student learns a subject pretty much in the order in which the subject developed over the ages. As a specific example, consider geometry. The earliest geometry may be called *subconscious geometry*, which originated in simple observations stemming from human ability to recognize physical form and to compare shapes and sizes. Geometry then became *scientific*, or *experimental, geometry,* and this phase of the subject arose when human intelligence was able to extract from a set of concrete geometrical relationships a general abstract relationship (a geometrical law) containing the former as particular cases. The bulk of pre-Hellenic geometry was of this experimental kind. Later, actually in the Greek period, geometry advanced to a higher stage and became *demonstrative geometry.* The basic pedagogical principle here under consideration claims, then, that geometry should first be presented to young children in its subconscious form, probably through simple artwork and simple observations of nature. Then, somewhat later, this subconscious basis is evolved into scientific geometry, wherein the pupils induce a considerable array of geometrical facts through experimentation with compasses and straightedge, with ruler and protractor, and with scissors and paste. Still later, when the student has become sufficiently sophisticated, geometry can be presented in its demonstrative, or deductive, form, and the advantages and disadvantages of the earlier inductive processes can be pointed out.

So we have here no quarrel with the pedagogical principle advocated by the tree of mathematics. But what about the tree itself? Does it still present a reasonably true picture of present-day mathematics? We think not. A tree of mathematics is clearly a function of time. The oak tree described above certainly could not, for example, have been the tree of mathematics during the great Alexandrian period. The oak tree does represent fairly well the situation in mathematics in the eighteenth century and a good part of the nineteenth century, for in those years the chief mathematical endeavors were the development, extension, and application of the calculus. But with the enormous growth of mathematics in the twentieth century, the general picture of mathematics as given by the oak tree no longer holds. It is perhaps quite correct to say that today the larger part of mathematics has no, or very little, connection with the calculus and its extensions. Consider, for example, the vast areas covered by abstract algebra, finite mathematics, set theory, combinatorics, mathematical logic, axiomatics, nonanalytical number theory, postulational studies of geometry, finite geometries, and on and on.

We must redraw the tree of mathematics if it is to represent mathematics of today. Fortunately there is an ideal tree for this new representation—the banyan tree. A banyan tree is a many-trunked tree, ever growing newer and newer trunks. Thus, from a branch of a banyan tree, a threadlike growth extends itself downward until it reaches the ground. There it takes root and over the succeeding years the thread becomes thicker and stronger, and in time becomes itself a trunk with many branches, each dropping threadlike growths to the ground.

There are some banyan trees in the world having many scores of trunks, and covering many city blocks in area. Like the great oak tree, these trees are both beautiful and long-lived; it is claimed that the banyan tree in India, against which Buddha rested while meditating, is still living and growing. We have, then, in the banyan tree a worthy and more accurate tree of mathematics for today. Over future years newer trunks will emerge, and some of the older trunks may atrophy and die away. Different students can select different trunks of the tree to ascend, each student first studying the foundations covered by the roots of his chosen trunk. All these trunks, of course, are connected overhead by the intricate branch system of the tree. The calculus trunk is still alive and doing well, but there is also, for example, a linear algebra trunk, a mathematical logic trunk, and others.

Mathematics has become so extensive that today one can be a very productive and creative mathematician and yet have scarcely any knowledge of the calculus and its extensions. We who teach mathematics in the colleges today are probably doing a disservice to some of our mathematics students by insisting that *all* students must first ascend the calculus trunk of the tree of mathematics. In spite of the great fascination and beauty of calculus, not all students of mathematics find it their "cup of tea." By forcing *all* students up the calculus trunk, we may well be killing off some potentially able mathematicians of the noncalculus fields. In short, it is perhaps time to adjust our mathematical pedagogy to fit a tree of mathematics that better reflects the recent historical development of the subject.

159° *Another analogy.*   The game of chess has always fascinated mathematicians, and there is reason to suppose that the possession of great powers of playing that game is in many features very much like the possession of great mathematical ability. There are the different pieces to learn, the pawns, the knights, the bishops, the castles, and the queen and the king. The board possesses certain possible combinations of squares, as in rows, diagonals, etc. The pieces are subject to certain rules by which their motions are governed, and there are other rules governing the players . . . One has only to increase the number of pieces, to enlarge the field of the board, and to produce new rules which are to govern either the pieces or the player, to have a pretty good idea of what mathematics consists.—J. B. SHAW

160° *Stepping down.*   Occasionally a college professor is appointed to the position of a deanship, and thenceforth no longer does anything academically productive. This has happened frequently to mathematics professors, and one is reminded of the following perhaps apocryphal story.

A mathematics professor developed a brain tumor and had to undergo an emergency operation. The surgeon detached the professor's cranium, took out the brain, and laid it on the table.

Right then a colleague of the professor arrived and announced that the professor had just been appointed to a deanship. With a whoop of joy, the professor bounced up from the operating table, slapped on his cranium, and dashed for the nearest exit. "Wait," cried the surgeon, "you've forgotten to put back your brain."

"I won't need it now," called back the professor over his shoulder, "I'm a dean."

161° *A least upper bound.*   S. Jennings, of the University of British Columbia, had become increasingly active in university administration. He was characterized by a colleague as "a least upper bound of non-deans."—LEO MOSER

162° *A great mistake.*   In 1688 Cambridge University selected Isaac Newton as their representative in Parliament. Not a very good choice, it would seem. During his entire tenure in Parliament, Newton's only known speech was a request to have a window opened.

PROFESSORS, TEACHERS, AND STUDENTS

163° *Julian Lowell Coolidge.*   Julian Lowell Coolidge, the great geometer at Harvard in the first half of the twentieth century, was a wit and a humorist in class. "I definitely try," it is said he once remarked, "when I teach, to make the students laugh. And while their mouths are open, I put something in for them to chew on."

164° *A special course.*   A mathematics professor was asked by his dean to prepare a special make-up course for a group of sick students at the university. The professor labeled the course, "Mathematics for Ill Literates."

165° *For the lazy student.*   Bennett Cerf has reported that Frank Boyden, a famous headmaster of Deerfield Academy, kept the following little poem on hand for lazy students:

> You can't go far just by wishing
> Nor by sitting around to wait.
> The good Lord provided the fishing—
> But you have to dig the bait.

One is reminded of the little jingle given in Item 190° of *Mathematical Circles Revisited.*

77

166° *A ticklish situation.*   A mathematics professor forgot an important afternoon appointment at his campus office that he had made with one of his students. After waiting at the professor's office for over a half hour, the student, with some trepidation, got up the courage to go to the professor's home in town. When he rang the bell, the door was answered by the professor's small son.

"Where can I get hold of your father?" inquired the nervous student.

"I wouldn't know," replied the little boy, "He's ticklish all over."

167° *Mars.*   The professor of mathematical astronomy, wishing to demonstrate something about the relative positions of some of the planets, said, "I will let my hat here represent the planet Mars. Any questions?"

"Yes," replied a student, "Is Mars inhabited?"

168° *Math exams.*   "A fool," sighed the annoyed mathematics professor, "can ask more questions in a few minutes than a wise man can answer in hours." At that, one of his students was heard to murmur in a barely audible voice, "No wonder so many of us flunked your last exam."

169° *Multiple use.*   On the wall in the hallway just outside the mathematics department lounge there is a row of hooks along with a sign reading, "For Faculty Members Only." Some witty student added in pencil below, "May also be used for hats and coats."

170° *How's that!*   A school board advertised for a teacher of algebra and geometry. One of the replies read: "Gentlemen, I noticed your advertisement for a teacher of algebra and geometry, male or female. Having been both for several years, I offer you my services."—LEO MOSER

171° *Research versus teaching.* A King once decided to honor that one of his subjects who had contributed the most to the furtherance of knowledge. Forthwith appeared a number of top-flight researchers in the various fields of study. After carefully considering the credentials of the candidates, the King noticed a stooped and shabbily dressed old woman standing in the background.

"Who is that woman?" asked the King.

"She has come merely to observe, Sire," replied the King's minister. "You see, she is interested in the outcome because she taught all these candidates when they were young."

The King descended from his throne and placed the wreath of honor on the old teacher's brow.

172° *Researchers and teachers.* Edith Wharton has pointed out, "There are two ways of spreading light: to be the candle, or the mirror that reflects it."

LECTURES

A LARGE part of a mathematics professor's life is spent in lecturing, either before college classes or at special mathematical gatherings. Sometimes amusing incidents occur during, or relative to, these lectures. Here are eight such stories gathered by Leo Moser. Note the similarity of the last three stories, wherein the existence of a *proof* of a mathematical result renders *checking* the result both mundane and extraneous.

173° *An understanding evaluation.* At the end of an hour lecture to a large audience at a mathematical congress in New York, C. S. Peirce remarked, "Come to think of it, there is only one person who might have had some chance of understanding the gist of my lecture today—and he is in South America."

—LEO MOSER

79

174° *A crackpot.*   Jacques Hadamard once planned a lecture trip in the United States. He wanted to meet American mathematicians and speak to many of the mathematics faculties concerning mathematical problems. He wrote, among other places, to Syracuse University. Since the head of the department was away, the letter reached the president of the university. The president refused Hadamard's request to have a lecture arranged for him. He argued that since Hadamard had not mentioned monetary remuneration, he must be a crackpot.—LEO MOSER

175° *Understanding.*   One year L. Infeld and R. Brauer gave a joint course on group theory and quantum mechanics. According to the graduate students present, Brauer didn't understand Infeld, Infeld didn't understand Brauer, and the students, of course, didn't understand either one.—LEO MOSER

176° *Interruption.*   Antoni Zygmund was constantly annoyed by a young female student who often interrupted his lecture with, "I don't understand this." Finally Zygmund said to her, "Don't worry about it. You are still so young."—LEO MOSER

177° *An uneasy start.*   Professor K. D. Fryer, of the University of Waterloo, was to give a lecture on continuous geometry at Queens University. Before the lecture, he filled the blackboard with very complicated-looking formulas and left a large note over these formulas: DO NOT ERASE. When the audience arrived they were rather worried by the complexity of the formulas and felt that the lecturer would soon leave them far behind. Fryer began the lecture with, "Well, I am sure we can dispense with all this." And he erased all the formulas.—LEO MOSER

178° *Checking.*   In a mathematics colloquium lecture at the University of Alberta, Max Wyman gave a long and complicated proof of a theorem which, once proved, could be checked quite easily. P. G. Rooney asked Wyman how the check came out and Wyman replied, "I didn't check it. I didn't have to. I proved it!" —LEO MOSER

179° *Bier Bauch.*   L. Bieberbach was overly fond of beer and as a result had a pot belly. He was known to his students as Bier Bauch (beer belly).

Bieberbach often had considerable difficulty with elementary arithmetic. His students knew his failing, so whenever he completed the proof of a theorem that could be illustrated by a numerical example, they would ask him to give such an example. After several unsuccessful attempts, he would give up in exasperation and exclaim, "Es ist doch so wie so bewiessen [In any case it has been proved]."—LEO MOSER

180° *Artin on graphs.*   Emil Artin gave a lecture on graphs at the University of Toronto. The problem he treated and solved was suggested by the question of how many chemical compounds satisfying certain valency conditions are possible. After the lecture, a chemist asked Artin how his result compared with the actual number of compounds. Artin replied, "I don't know and I don't care."—LEO MOSER

QUADRANT THREE

*From author's jokes
to a difficult problem*

AUTHORS AND BOOKS

SOMETIMES an author of a mathematics text will indulge in a literary prank or spring a joke of significance only to the brethren. And then again, sometimes a nonmathematical author will attempt a bit of mathematical discussion that a mathematician may find interesting or amusing. Here are some examples.*

181° *Authors' jokes.*　Leonard Gillman and Robert McDowell, in their book *Calculus* (W. W. Norton, 1973), include, "for the sake of completeness," a theorem concerning greatest lower bounds. Elsewhere in the book they state that "a little reflection shows that the graph of f^{-1} is the mirror image . . . of the graph of f."

182° *The Chinese Remainder Theorem.*　On p. 72 of Harold M. Stark's *An Introduction to Number Theory* appears, in connection with the Chinese Remainder Theorem, the following cute footnote: "More rarely known as the Formosa Theorem."

183° *Mohr–Mascheroni constructions.*　Commenting on Mohr–Mascheroni constructions (constructions utilizing the compasses only), M. H. Greenblatt, in his book *Mathematical Entertainments,* says: "Mohr of these Mascheroni constructions are described in a booklet by A. N. Kostovskii."

184° *Honoring a friend.*　There is a story that says when the witty Paul Halmos wrote his little gem, *Finite-Dimensional Vector Spaces,* he promised a friend that he would mention the friend's name in the book. True enough, in the index of the book one finds the name, "Hochschild, G. P." The page reference is 198, which

---

*Earlier examples may be found in Item 192° of *Mathematical Circles Revisited* and Items 344°, 345°, 346°, and 347° of *Mathematical Circles Squared.*

turns out to be the number of the page in the index where the name is mentioned.

Hochschild later became an eminent professor of mathematics at the University of California at Berkeley.

185° *Close.*   In the book *Introduction to Modern Prime Number Theory,* a certain theorem is proved by the indirect method. The contradiction comes out in the form $333 \geq 1000/3$. The author, T. Estermann, remarks: "The narrow margin by which this contradiction was obtained reminds me of the story of the Scotsman who looked suspiciously at his change, and when asked if it was not enough, said: 'Yes, but only just.' "—LEO MOSER

186° *Among the authors.*   Helmut Hasse, in the index of authors at the rear of his *Vorlesungen über Zahlentheorie* (1950), lists "Gott," with a reference to page 1. On page 1 one finds quoted Kronecker's famous remark, "Die ganzen Zahlen hat Gott gemacht, alles andere ist Menschenwerk."

187° *A deletion.*   In my Springer textbook *Vorlesungen über Zahlentheorie,* at the very beginning, I mention Kronecker's famous dictum and Dedekind's quite opposite opinion. In 1953 a Russian translation of this book appeared in Moscow, and in it the first paragraph is left out. When I went to Moscow in 1963, they told me there that this paragraph was not admitted by the State Editors of Foreign Literature, because God does not exist.

—HELMUT HASSE

188° *Whewell's poetry.*   Dr. William Whewell, when Master of Trinity College, much to the amusement of his students, unintentionally rhymed some of the prose in the first edition (1819) of his *An Elementary Treatise on Mechanics.* He wrote: "There is no force, however great, can stretch a cord, however fine, into a horizontal line, which is accurately straight," which forms the tetrastich:

There is no force, however great,
Can stretch a cord, however fine,
Into a horizontal line,
Which is accurately straight.

The accidental rhyme was first brought to Whewell's attention when Professor Sedgwick of the geology department at Cambridge recited the lines in an after-dinner speech. Whewell, who failed to see any humor in the matter, altered the lines in the following edition of his work so as to eliminate the poem. Ironically, during his lifetime Whewell published two books of serious poetry, but the lines quoted above constitute the only "poem" by him remembered today.

189° *A criticism.*   David Widder, of Harvard University, was once asked, "What is your opinion of Bateman's book on differential equations? It is a very good book, is it not?" "That book!" Widder exclaimed, "Open it to any page at random and show me any statement in heavy print and I will give you a counterexample."—LEO MOSER

190° *Proving a postulate.*   At a social gathering given by the von Rinnlingens in Thomas Mann's famous short story *Little Herr Friedemann,* a number of the guests retire to the smoking-room. We read:

> Some of the men stood talking in this room, and at the right of the door a little knot had formed round a small table, the center of which was the mathematics student, who was eagerly talking. He had made the assertion that one could draw through a given point more than one parallel to a straight line; Frau Hagenström had cried out that this was impossible, and he had gone on to prove it so conclusively that his hearers were constrained to behave as though they understood.

Thomas Mann was born in Lübeck, Germany, in 1875, and died in Zurich, Switzerland, in 1955. *Little Herr Friedemann* was his first literary success; he wrote it in his early twenties while on a two-year stay in Italy.

87

191° *A classification of minds.* I was just going to say, when I was interrupted, that one of the many ways of classifying minds is under the heads of arithmetical and algebraical intellects. All economical and practical wisdom is an extension of the following arithmetical formula: $2 + 2 = 4$. Every philosophical proposition has the more general character of the expression $a + b = c$. We are mere operatives, empirics, and egotists until we learn to think in letters instead of figures.

—OLIVER WENDELL HOLMES
The Autocrat of the Breakfast Table

192° *About vectors.* It is rumored that there was once a tribe of Indians who believed that arrows are vectors. To shoot a deer due northeast, they did not aim an arrow in the northeasterly direction; they sent two arrows simultaneously, one due north and the other due east, relying on the powerful resultant of the two arrows to kill the deer.

Skeptical scientists have doubted the truth of this rumor, pointing out that not the slightest trace of the tribe has ever been found. But the complete disappearance of the tribe through starvation is precisely what one would expect under the circumstances; and since the theory that the tribe existed confirms two such diverse things as the NONVECTORIAL BEHAVIOR OF ARROWS and the DARWINIAN PRINCIPLE OF NATURAL SELECTION, it is surely not a theory to be dismissed lightly.

—BANESH HOFFMAN
About Vectors (Prentice-Hall, Inc., 1966)

193° *Asymptotic approach.* . . . he seemed to approach the grave as an hyperbolic curve approaches a line—less directly as he got nearer, till it was doubtful if he would ever reach it at all.

THOMAS HARDY
Far from the Madding Crowd
about the malster, in Chapter XV

194° *A literary allusion.* James Hilton, in his haunting novel *Lost Horizon,* describes Conway's first impression of the foreground leading to Shangri-La, and of the more remote, conical, snow-covered Mt. Karakal that tenuously towers over Shangri-La.

> It was not a friendly picture, but to Conway, as he surveyed, there came a queer perception of fineness in it, of something that had no romantic appeal at all, but a steely, almost intellectual quality. The white pyramid in the distance compelled the mind's assent as passionlessly as a Euclidean theorem . . . .

Later in the story we read of the interplay of the atmosphere and the mystery of Shangri-La on Conway:

> Its atmosphere soothed while its mystery stimulated, and the total sensation was agreeable. For some days now he had been reaching, gradually and tentatively, a curious conclusion about the lamasery and its inhabitants; his brain was still busy with it, though in a deeper sense he was unperturbed. He was like a mathematician with an abstruse problem—worrying over it, but worrying very calmly and impersonally.

At still another point in the story, Conway notes his reaction to the discourses of the High Lama:

> At times he had the sensation of being completely bewitched by the mastery of that central intelligence, and then, over the little pale blue tea-bowls, the celebration would contract into a liveliness so gentle and miniature that he had an impression of a theorem dissolving limpidly into a sonnet.

One is reminded of Sherlock Holmes' disparaging comment to Watson on the completion of the latter's first book, *A Study in Scarlet*:

> Honestly, I cannot congratulate you on it. Detection is, or ought to be, an exact science, and should be treated in the same cold and unemotional manner. You have attempted to tinge it with romanticism, which produces the same effect as if you worked a love story or elopement into the 5th proposition of Euclid.

DEFINITIONS

THE forming of accurate definitions is of great importance in mathematics.

195° *A definition.*   A little girl was asked for her definition of *nothing.* *"Nothing,"* she replied, "is like a balloon with its skin off."

196° *On definitions.*   Making good definitions is not easy, The story goes that when the philosopher Plato defined *man* as "a two-legged animal without feathers," Diogenes produced a plucked cock and said, "Here is Plato's man." Because of this, the definition was patched up by adding the phrase "and having broad nails"; and there, unfortunately, the story ends. But what if Diogenes had countered by presenting Plato with the feathers he had plucked?

—BANESH HOFFMAN
About Vectors (Prentice-Hall, Inc., 1966)

197° *A suggestion.*   A short word to replace the often-used phrase "transformation of coordinates" would be handy in geometry. How about "code"?

198° *Clayton Dodge's rectangle (wrecked angle).*

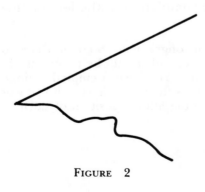

FIGURE   2

90

199° *Cute.* Question: Define *acute right triangle.* Answer: A right triangle with pretty legs.—CLAYTON W. DODGE

200° *A Fourier series.* Yea + Yea + Yea + Yea.
—CLAYTON W. DODGE

201° *Yes.* On his preliminary Ph.D. oral examination, W. Crawford was asked to explain the meaning of the terms *homeomorphism* and *homomorphism.* He knew what a homeomorphism is, but couldn't quite remember the definition of a homomorphism. He nevertheless concocted a correct answer, namely: "A homomorphism is an isomorphism that didn't quite make it."

—LEO MOSER

LOGIC

A MATHEMATICAL theory results from the interplay of two factors, a set of postulates and a logic. The set of postulates constitutes the basis from which the theory starts and the logic constitutes the rules by which such a basis may be expanded into a body of theorems. [From Howard Eves and C. V. Newsom, *An Introduction to the Foundations and Fundamental Concepts of Mathematics,* revised edition. Holt, Rinehart and Winston, 1965.]

202° *Cause and effect.* It is well known that for every cause there is an effect and for every effect there is a cause, but sometimes the two are so inextricably tangled together that it is difficult to tell which is which. Consider, for example, the case of the two desert nomads visiting Wisconsin for their first time and seeing a speedboat pulling a zig-zagging water skier about a lake. "Why does the boat go so fast?" inquired one of the nomads. Replied the other, "Because it is chased by a lunatic on a string."

203° *Deduction.* Some postal authorities are amazingly persevering and ingenious. It is said that one such official came upon a letter addressed to:

Wood.
Mr.
Mass.

and promptly and correctly had it delivered to:

Mr. Underwood
Andover,
Mass.

204° *No wonder.*   Two Irish ladies argued with one another every day from their windows, across an intervening lane. They never agreed, which was to be expected of course, for they argued from different premises.

205° *Benchley's law of dichotomy.*   Robert Benchley remarked, "There may be said to be two classes of people in the world: those who constantly divide the people of the world into two classes and those who do not."

206° *So it seems.*   While watching a TV newscast, a woman turned toward her husband and observed: "It seems to me that the majority of people in this country belong to some minority group."

207° *No doubt.*   A reporter asked Undersecretary of State Robert J. McCloskey to comment on the art of decision-making. The Undersecretary replied: "One thing you must keep in mind here: that all decisions aren't made until all decisions are made."

208° *Logical.*   Mayor Richard J. Daly of Chicago, when asked by a reporter to comment on the current trucking strike, replied: "What keeps people apart is their inability to get together."

209° *Necessary and sufficient.*   A local radio announcer, commenting on hazardous driving conditions, advised: "Please don't do any unnecessary driving unless it's absolutely necessary."

210° *Come again.*　An auto-accident report to an insurance company contained the following statement: "My car sustained no damage whatever and the other car somewhat less."

211° *Separate or together.*　Finkbeiner and I were having breakfast in San Antonio and asked for the check. The waitress said, "You want them separate or together?" I said, "Separate, please." She turned to Finkbeiner and said, "You want yours separate too?"—RALPH BOAS

212° *Who's who?*　I met Tom Jones and I said, "How have you been, Jones?" And Jones replied, "Fair to middling, thank you. How have you been, Smith?" "Smith," I said, "That's not my name!" "Nor is my name Jones," said the other fellow. Then we looked each other over again, and true! It was neither of us!

213° *Proof.*　Ambling Andy was telling a group about his great love of walking. He said that as he kept extending his daily walks, he found himself farther and farther from home each day when evening approached. He solved the difficulty by moving to a house halfway up a big conical hill, and thenceforth he spent his days walking round and round the side of the hill, starting each day with the rising sun behind him and ending each day with the setting sun before him. After some years of this he made the tragic discovery that by continual walking around the conical hill he had worn his uphill leg shorter than his downhill leg, and his friends commiserated that the walking days of Ambling Andy were essentially over. But, no. Andy merely reversed his direction of walking around the hill until he wore his long leg down to the length of his short leg. When a listener appeared incredulous, Andy proved the whole matter by standing before the disbelieving person and pointing out: "See, aren't my two legs the same length?"

214° *Of course.*　Question: How do you pronounce the word "ghot"?

Answer: As "fish," by taking the "gh" as in "laugh," the "o" as in "women," and the "t" as in "nation."

215° *A syllogism.*

> One cat has one more tail than no cat.
> No cat has two tails.
> _____
> Therefore, one cat has three tails.

216° *Double negative.*   A five-year old was being cajoled to recite a poem. To prod him, his mother said, "You just don't know, do you?" To this the youngster replied, "I do not don't know!"

217° *Naturally.*   Grandma can never find her glasses any-more—so she drinks from the can.

218° *An exception.*   An exception is never used to prove a rule, but rather to test a rule. That is, if a rule works well in exceptional (rather than simple) cases, the chances are excellent that it is a good rule.

219° *Ability versus nonability.*

(1) I love mathematics; I attend the meetings of the Mathematics Club and I study the subject a great deal. Still I can't do well in mathematics courses.

(1) I love baseball; I watch it on TV and I study books on the subject. Still I'm a pretty awful player.

(2) I study mathematics much harder than Joe does, but he still gets much better grades.

(2) I practice wrestling much more than Joe does, but he still always beats me.

220° *A law of physics.*   In a little New England town, a brass band of a dozen musicians was blaring away, under the energetic leadership of a puffing and uniformed maestro, before one of the houses of the town.

A curious passing tourist stopped and inquired of an on-looker, "Who are they serenading?"

"Oh, the mayor," was the reply. "Why doesn't he come to the

94

door or a window and acknowledge the compliment?" continued the tourist.

"Because that's the mayor leading the band," explained the onlooker. "You can't expect him to be in two different places at the same time, can you?"

221° *Language difficulties.* A Japanese mathematician wrote to an American mathematician requesting a reprint. He concluded with: "And above all, please excuse my shameless desire."

An Indian student wrote the head of the mathematics department of a Canadian university and expressed interest in continuing his studies in Canada. His financial condition apparently was not good, so he inquired, "I wonder what you can do to help me make both my ends meet."—LEO MOSER

ON MATHEMATICS AND LOGIC

MANY eminent, and some not so eminent, persons have commented on the relative positions of logic and mathematics.

222° *Bacon's misconceived notion.* It has come to pass, I know not how, that Mathematics and Logic, which ought to be but the handmaids of Physic, nevertheless presume on the strength of the certainty which they possess to exercise dominion over it.

—FRANCIS BACON
De Augmentis

223° *Successive approximation.* It is commonly considered that mathematics owes its certainty to its reliance on the immutable principles of formal logic. This . . . is only half the truth imperfectly expressed. The other half would be that the principles of formal logic owe such a degree of permanence as they have largely to the fact that they have been tempered by long and varied use by mathematicians. "A vicious circle!" you will perhaps say. I should rather describe it as an example of the process known by mathematicians as the method of successive approximation.—MAXIME BÔCHER

224° *Which is which?*   George Bruce Halsted has described mathematics as the "giant pincers" of logic. Isn't it equally true that logic is the "giant pincers" of mathematics?

225° *An identity.*   . . . the two great components of the critical movement, though distinct in origin and following separate paths, are found to converge at last in the thesis: Symbolic Logic is Mathematics, Mathematics is Symbolic Logic, the twain are one. —CASSIUS J. KEYSER

226° *Monadology.*   It seemed to Leibniz that if all the complex and apparently disconnected ideas which make up our knowledge could be analysed into their simple elements, and if these elements could each be represented by a definite sign, we should have a kind of "alphabet of human thoughts." By the combination of these signs (letters of the alphabet of thought) a system of true knowledge would be built up, in which reality would be more and more adequately represented or symbolized. . . . Thus it seemed to Leibniz that a synthetic calculus, based upon a thorough analysis, would be the most effective instrument of knowledge that could be devised. "I feel," he says, "that controversies can never be finished, nor silence imposed upon the Sects, unless we give up complicated reasonings in favor of simple *calculations,* words of vague and uncertain meaning in favor of fixed symbols." Thus it will appear that "every paralogism is nothing but *an error of calculation.*" "When controversies arise, there will be no more necessity of disputation between two philosophers than between two accountants. Nothing will be needed but that they should take pen in hand, sit down with their counting-tables, and (having summoned a friend, if they like) say to one another: *Let us calculate.*" —ROBERT LATTA

227° *The foundation for a scientific education.*   Formal thought, consciously recognized as such, is the means of all exact knowledge; and a correct understanding of the main formal sciences, Logic and Mathematics, is the proper and only safe foundation for a scientific education.—ARTHUR LEFEVRE

228° *The identity again.*   Pure mathematics was discovered by Boole in a work he called 'The Laws of Thought' . . . His work was concerned with formal logic, and this is the same thing as mathematics.—BERTRAND RUSSELL

229° *And again.*   Mathematics is but the higher development of Symbolic Logic.—W. C. D. WHETHAM

230° *Symbolic Logic.*   Symbolic Logic has been disowned by many logicians on the plea that its interest is mathematical, and by many mathematicians on the plea that its interest is logical.
—ALFRED NORTH WHITEHEAD

231° *New logics.*   [This, and the succeeding, items are adapted from Howard Eves and C. V. Newsom, *An Introduction to the Foundations and Fundamental Concepts of Mathematics,* revised edition. Holt, Rinehart and Winston, 1965.]
An interesting analogy (if it is not pushed too far) exists between the parallelogram law of forces and the postulational method. By the parallelogram law, two component forces are combined into a single resultant force. Different resultant forces are obtained by varying one or both of the component forces, although it is possible to obtain the same resultant force by taking different pairs of initial component forces. Now, just as the resultant force is determined by the two initial component forces, so (see Figure 3) is a mathematical theory determined by a set of postu-

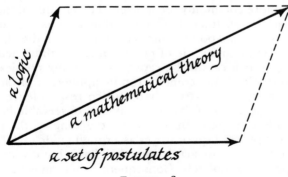

FIGURE   3

lates and a logic. That is, the set of statements constituting a mathematical theory results from the interplay of an initial set of statements, called the postulates, and another initial set of statements, called the logic or the rules of procedure. For some time mathematicians have been aware of the variability of the first set of initial statements, namely, the postulates, but until very recent times the second set of initial statements, namely, the logic, was universally thought to be fixed, absolute, and immutable. Indeed, this is still the prevailing view among most people, for it seems quite inconceivable, except to the very few students of the subject, that there could be any alternative to the laws of logic stated by Aristotle in the fourth century B.C. The general feeling is that these laws are in some way attributes of the structure of the universe and that they are inherent in the very nature of human reasoning. As with many other absolutes of the past, this one, too, has toppled, but only as late as 1921. The modern viewpoint can hardly be more neatly put than in the following words of the outstanding American logician, Alonzo Church.

> We do not attach any character of uniqueness or absolute truth to any particular system of logic. The entities of formal logic are abstractions, invented because of their use in describing and systematizing facts of experience or observation, and their properties, determined in rough outline by this intended use, depend for their exact character on the arbitrary choice of the inventor. We may draw the analogy of a three-dimensional geometry used in describing physical space, a case for which, we believe, the presence of such a situation is more commonly recognized. The entities of the geometry are clearly of abstract character, numbering as they do planes without thickness and points which cover no area in the plane, point sets containing an infinitude of points, lines of infinite length, and other things which cannot be reproduced in any physical experiment. Nevertheless the geometry can be applied to physical space in such a way that an extremely useful correspondence is set up between the theorems of the geometry and observable facts about material bodies in space. In building the geometry, the proposed application to physical space serves as a rough guide in determining what properties the abstract entities shall have, but does not assign these properties completely. Consequently there may be, and actually are, more than one geometry whose use is feasible in describing

physical space. Similarly, there exist, undoubtedly, more than one formal system whose use as a logic is feasible, and of these systems one may be more pleasing or more convenient than another, but it cannot be said that one is right and the other wrong.

232° *A brief history of the new logics.* New geometries first came about through the denial of Euclid's parallel postulate, and new algebras first came about through the denial of the commutative law of multiplication. In a similar fashion, the new so-called "many-valued" logics first came about by denying Aristotle's law of excluded middle. According to this law, the disjunctive proposition $p \lor$ not-p is a tautology, and a proposition p in Aristotelian logic is always either true or false. Because a proposition may possess any one of two possible truth values, namely truth or falsity, this logic is known as a two-valued logic. In 1921, in a short two-page paper, J. Lukasiewicz considered a three-valued logic, or a logic in which a proposition p may possess any one of three possible truth values. Very shortly after, and independently of Lukasiewicz's work, E. L. Post considered m-valued logics, in which a proposition p may possess any one of m possible truth values, where m is an integer greater than 1. If m exceeds 2, the logic is said to be *many-valued.* Another study of m-valued logics was given in 1930 by Lukasiewicz and A. Tarski. Then, in 1932, the m-valued truth systems were extended by H. Reichenbach to an infinite-valued logic, in which a proposition p may assume any one of infinitely many possible values.

Not all new logics are of the type just discussed. Thus A. Heyting has developed a symbolic two-valued logic to serve the intuitionist school of mathematicians; it differs from Aristotelian logic in that it does not universally accept the law of excluded middle or the law of double negation. Like the many-valued logics, then, this special-purpose logic exhibits differences from Aristotelian laws. Such logics are known as *non-Aristotelian logics.*

Like the non-Euclidean geometries, the non-Aristotelian logics have proved not to be barren of application. Reichenbach actually devised his infinite-valued logic to serve as a basis for the mathematical theory of probability. And in 1933 F. Zwicky observed that many-valued logics can be applied to the quantum theory of modern physics. Many of the details of such an application have been supplied by Garrett Birkhoff, J. von Neumann, and H. Reichenbach. Lukasiewicz has employed three-valued logics to establish the independence of the postulates of familiar two-valued logic. The part that non-Aristotelian logics may play in the future development of mathematics is uncertain but intriguing to contemplate; the application of Heyting's symbolic logic to intuitionist mathematics indicates that the new logics may be mathematically valuable.

COUNTING

THE basis of counting is the notion of a one-to-one correspondence.

233° *A one-to-one correspondence.* One of the earliest literary references to the primitive method of keeping a count by setting up a one-to-one correspondence occurs in the Homeric legends about Ulysses. When Ulysses left the land of the Cyclops, after blinding the one-eyed giant Polyphemus, it is narrated that that unfortunate old giant would sit each morning near the entrance to his cave and from a heap of pebbles would pick up one for each ewe that he let pass out of the cave. Then, in the evening, when the ewes returned, he would drop one pebble for each ewe that he admitted to the cave. In this way, by exhausting the supply of pebbles he had picked up in the morning, he was assured that all his flock had returned in the evening.

234° *Another one-to-one correspondence.* Like the famous David Hilbert in his older age, there was a retired professor of mathematics who was quite a devil with the ladies, still charming the daylights out of them at seventy-seven. In fact, on his seventy-seventh birthday, the professor decided to cut a notch in his cane for each of his new conquests. And that's what killed him on his seventy-eight birthday; he made the mistake of leaning on his cane.

235° *Buttons.* Carrier: "Can 'ee spell?"
"Yes!"

Carrier: "Cipher?"
"Yes!"

Carrier: "That's more than I can; I counts upon my fingers. When they be used up, I begin upon my buttons. I han't got no buttons—visible, that is—'pon my week-a-day clothes, so I keeps the long sums for Sundays, and adds 'em up and down my weskit during sermon. Don't tell any person."
"I won't."

Carrier: "That's right; I don't want it known. Ever seen a gypsy?"
"Oh, yes, often."

Carrier: "Next time you see one, you'll know why he wears so many buttons. You've a lot to learn."

—SIR ARTHUR QUILLER-COUCH
The Ship of Stars

NUMBERS

HERE are still more stories—some silly, some serious—about numbers.

236° *Nein.* In view of the recent revealing of hanky-panky conducted by a number of our government officials in Washington, D.C., an attractive German fraulein in that city was questioned if she had ever been away on trips with senators. She indignantly replied, *"Nein"*—and so was deported.

237° *Round letters.* And then there was the boss who complained that his secretary spelled like a mathematician—she rounded her words off to the nearest letter of the alphabet.

238° *Experimental error.* An experimental physicist was testing the conjecture that all odd numbers are prime. He started by saying, "Three—that's prime. Five—that's prime, too. Seven—

still prime. Nine—oops! What's wrong? But let's go on a bit further. Eleven—that's prime. Thirteen—still prime. Nine must have been an experimental error!"—M. H. GREENBLATT

239° *The psychology of numbers.*   There was a famous Betty Crocker packaged cake mix where, on the box, the busy cook was instructed to add to the contents of the package one egg and half a cup of water *less* one tablespoon, mix for two minutes at medium speed with an electric mixer, pour into a greased 9-inch pan, and then bake for 35 minutes in a moderate (350°F) oven.

Now the cake would bake just as well using a full half-cup of water, but interviews and consumer research revealed that then the recipe would appear too simple, and would leave the baker with guilt feelings. That is, if the cake batter should be too easy to prepare, the cook's basic honesty would prevent her from believing that she had made the cake herself, and her pride in the finished cake would be destroyed. So the recipe had to be altered, still keeping it very simple, yet creating an aura of labor, measurement, and finicky detail. It turned out that asking the cook to remove one tablespoon of water was precisely the right psychological fillip, and was well within the tolerance of error when using ordinary household measuring devices. The cook's guilt feelings were removed, her feeling of having been creative was bolstered, and the recipe sold millions upon millions of packages of the cake mix.

240° *The ten digits in the smallest Pythagorean triangle.*

1—The inradius
2—The indiameter, $a + b - c$
3—The short leg, a
4—The long leg, b
5—The hypotenuse, c
6—The area, $ab/2$
7—The sum of the legs, $a + b$

8—The short leg plus the hypotenuse, $a + c$
9—The long leg plus the hypotenuse, $b + c$
0—The esoteric significance of it all

—CHARLES W. TRIGG

241° *A three-part "catch" by Henry Purcell, 1731.*

When V and I together meet
We make up 6 in House or Street,

Yet I and V may meet once more
And then we 2 can make but 4,

But when that V from I am gone
Alas poor I can make but one.

[From "The Catch Club or Merry Companions, being a Choice Collection of the Most Diverting Catches for Three and Four Voices," Part 1. Reprint, Da Capo Press, N.Y., 1965.]

Originally, a catch was a round for three or more unaccompanied voices, written out as one continuous melody, each succeeding singer taking up a part in turn. Later, such a round on words combined with ludicrous effect. (From *Webster's New International Dictionary of the English Language,* second edition, unabridged, 1956.)—JOHN F. BOBALEK

242° *A curious permutation of digits.* The history of modern America began in 1492, the year the Italian navigator Christopher Columbus reached the New World. The atomic age began in 1942, the year the Italian physicist Enrico Fermi achieved the first nuclear reaction.

243° *Evaluating some desirable characteristics.* Jean Jacques Rousseau (1712–1778), the French writer and political philosopher whose ideas helped to inspire the leaders of the French Revolution, was once asked by a young lady, "What characteristics must a young lady have in order to make her man happy?" Rousseau wrote on a piece of paper:

> Beauty: 0
> Housekeeping ability: 0
> Wealth: 0
> Good nature: 1

He explained: "If a girl has nothing but good nature, she has 1. If she has other fine qualities she can be valued at 10, 100, 1000, etc. However, without the 1 in front, she is nothing."

244° *An interesting use for the new math.*   The following real estate ad appeared in *Down East* (the Magazine of Maine), April, 1973, p. 80.

> Prime Location.   Enjoy an equally lovely salt water view from almost every window all year round in this sturdily built Colonial. You will want to paint and paper, but it's priced accordingly. Charming fireplace, 1-½ baths, 4 bedrooms (or 5 if you've taken new math), garage, tool shed, low taxes.
>
> —JANET B. GOODHUE, INC.

245° *Misinterpretation.*   A man and his wife, both in their late 60s and retired, were shopping in a local supermarket. The wife stopped at the magazine rack, studied the display, selected a magazine and put it in her shopping basket. A few minutes later she returned to the rack and replaced the magazine. "Don't you want to buy that?" asked her husband.

"Not now," she replied. "I was intrigued by the title, 'Sex in the 70s.' But then it dawned on me, they mean the 1970s."

246° *An expression for forty*   One Malinke expression for "forty" is *dibi*, "a mattress," from the union of the forty digits, "since the husband and the wife lie on the same mattress and have a total of forty digits between them," to quote Delafosse.

—CLAUDIA ZASLAVSKY

247° *Rational digits again.* In Item 1° of *Mathematical Circles Revisited,* we presented some imagined, but unhistorical, explanations of the origin of our digit symbols. Figure 4 shows still another such "rational" explanation.

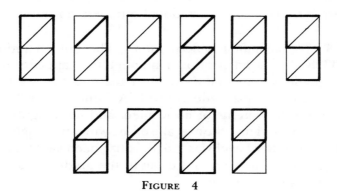

FIGURE 4

248° *In Rome.* On the popular TV program *Hollywood Squares,* Paul Linde was asked, "If you are in Rome and a woman comes up to you and says she is XL, how old is she?" Paul Linde's instantaneous reply was, "Who cares, she's extra large."

249° *Beasting the author.* John F. Bobalek submitted the following to English key: HOWARD W. EVES, A PROFESSOR OF MATHEMATICS AND DOCTOR OF PHILOSOPHY.

250° *Beasting Hitler.* One can easily beast *Hitler* by adding up the numerical values of the six letters, counting A as 100, B as 101, C as 102, and so on.

251° *An odd comment.* Upon hearing that there are no odd perfect numbers less than 10^{100}, Professor John Mairhuber commented: "That seems to indicate that odd perfect numbers are not abundant." To this, Reverend Thomas Vining remarked: "True, but if a perfect number *is* abundant, it is certainly odd."

252° *There are no irrational numbers.* R. F. Jolly has reported the following dialogue he once had with a student:

Student: Irrational numbers do not exist!

Jolly: Oh?

Student: God is a perfect rational being; therefore all of His works are rational. Hence, irrational numbers do not exist.

253° *The ubiquitous golden ratio.* The golden ratio, $\phi =$ 1.61803398 . . . is an irrational number that appears in many diverse situations. One of the strangest attempts to show the ever-present nature of ϕ was made by one F. A. Lonc, of New York City. He measured the heights of 65 women, and the heights of their navels. He reported that the average of the woman's height divided by the height of her navel was 1.618. It has been said that women whose measurements were not close to the golden ratio testified to hip injuries or other deformities.

LOGARITHMS

254° *A book of sensible numbers.* Once, when the German satirist Gottlieb Wilhelm Rabener (1714–1771) called on Abraham Gottlieb Kästner (1719–1800) at Göttingen, he found a table of logarithms on Kästner's desk. Thumbing through the table he commented, "There isn't a sensible word in the whole book." "No," replied Kästner, "but there are many sensible numbers."

255° *A change in education.* Kiano Fisher (1824–1907), the famous German philosopher and historian, has told an amusing story about logarithms. It seems that a couple of his schoolmates had a simple-minded uncle who would occasionally glance at their homework, even though he could not understand anything of what he saw. He once came upon the lads while they were using Vega's *Table of Logarithms,* a book which from cover to cover contains only numbers. His curiosity aroused, the uncle asked the boys as to the contents of the book, and they told him it contained all the house numbers in Europe. That evening, at the local beer garden, the uncle told some old friends of his that while there was a great deal to learn in school in his day, it was nothing compared to present demands—the children now had to learn all the house numbers of Europe.

256° *Graphomaths.*   There was once a popular novel in which the author created a mathematics professor who, as a student, had proved his mathematical qualifications by memorizing Vega's enormous table of logarithms, and as a result suffered a nervous breakdown.

Augustus De Morgan coined the term "graphomath" for a person who, ignorant of mathematics, attempts to describe a mathematician. Sir Walter Scott was a graphomath when he had his dreamy calculator, Dave Ramsay, swear "by the bones of the immortal Napier." Another graphomath had his mathematical hero constantly "doing sums," and still another claimed his hero was always engaged in "calculating double differentials."

257° *A sad and tragic end.*   Baron Georg von Vega (1754–1802) was a military officer and mathematician, famous for his military campaigns and his great table of logarithms. He was reportedly murdered in 1802 for "his money and his watch."

258° *An important work.*   The invention of logarithms and the calculation of the earlier tables form a very striking episode in the history of exact science, and, with the exception of the *Principia* of Newton, there is no mathematical work published in the country which has produced such important consequences, or to which so much interest attaches as to Napier's *Descriptio.*

—J. W. L. GLAISHER
Encyclopedia Britannica, 9th Edition

ARITHMETIC

259° *Cauchy and the calculating prodigy.*   The French shepherd boy and calculating prodigy, Henri Mondeux (b. 1826), gave a performance in 1840, when he was fourteen years old, before the students and faculty of the École Polytechnique. The performance started with questions from the audience. Mondeux solved the problems quickly and correctly. At length a problem was proposed that involved long calculations. Mondeux sat in deep concentration, mentally working on the problem. He seemed to be almost finished when a man in the audience stood up and triumphantly gave the answer. It was the mathematician Augustin-Louis Cauchy (1789–1857).

To protect the young prodigy, the physicist Gaspard Gustave de Coriolis (1792–1843), who was also present, challenged Cauchy to present the boy with a problem to solve. Cauchy asked the lad to find the sum of the fourth powers of the first twenty natural numbers. Mondeux closed his eyes and proceeded with the calculation. Cauchy also closed his eyes, but suddenly opened them and called out, "722,666." Cauchy, of course, had not followed the tedious path of his young rival. He used, instead, the formula

$$\sum_{i=1}^{n} i^4 = \frac{n(n + 1)(2n + 1)(3n^2 + 3n - 1)}{2 \cdot 3 \cdot 5}$$

Cauchy was able to give the desired result by computing the product of 574 and 1259, which he was able to do in his head rather easily.

—ELWOOD EDE
(freely translated from W. Ahrens, *Mathematiker-Anekdoten*)

260° *The Human Computer.*   Willem Klein (b. 1912) is known as "The Human Computer." As the son of a Jewish doctor living in Amsterdam during the war, he and the rest of the family had to go into hiding from the Nazis. It was during this period of hiding that Willem Klein discovered a passion for numbers, and by much practice became highly skilled at long mental calculations. After the war he made a living performing in European music halls as "The Man with the 10,000-Pound Brain." In 1958 CERN (European Center of Nuclear Research), located near Geneva, recognized his unusual talents and hired him, for at that time he was far more efficient than the center's computer.

Still very much of a showman, Klein recently set out to break his own record of 3 minutes and 43 seconds, succeeding by one full minute. Before a rapt audience at the CERN auditorium, Klein mentally computed the 73rd root of a 499-digit number. At the end of 2 minutes and 43 seconds he announced his answer, 6,789,235. His answer was confirmed by a modern electronic computer.

Since modern computers have overtaken Klein in speed, the great mental computer intends this year (1976) to retire from CERN, return to Amsterdam, visit the schools, and "Show children how to have fun with numbers."

261° *Gender numbers.* Adopting the Pythagorean convention that even numbers are "female" and odd numbers are "male," the following observations can be made:

1. All males are odd.
2. For any two males, there's a female (it just adds up!).
3. There's only one prime female.
4. Any female is twice a male.
5. When females start multiplying their numbers, a male can't get in edgewise.
6. Between any two females there is at least one male.
7. Any prime female has the potential to be a two-timer.
8. A male always follows a female and vice versa.
9. There is at least one prime factor in every female which is not a factor common to any male.
10. Taking all into consideration, when you get down to the roots of the matter, males and females are equally irrational.

—STEPHEN J. TURNER

262° *Oversensitivity.* It is said that Picasso, at the age of eleven, still could not do arithmetic, because the numeral 7 looked to him like a nose upside down.

263° *A good excuse.*   When the teacher asked Johnny where his arithmetic homework paper was, the boy explained: "On my way to school I made an airplane out of it, and someone hijacked it."

264° *Why not?*   Did you hear about the plant in the math teacher's room? It grew square roots.

265° *To tell the truth.*   On his arithmetic test, Johnny worked a problem three different ways and obtained three different answers. Perplexed as to which was the correct answer, he finally in exasperation muttered: "Will the real answer please stand up?"

266° *Close, indeed!*   In my arithmetic test this morning I was mighty close to the right answers. They were only two seats away.

267° *The day's accomplishment.*   "What did you learn in arithmetic today?" asked Johnny's mother.

"I learned that seven and seven makes fifteen," Johnny replied.

"But that's not right," said his mother.

"Oh? Well then I didn't learn anything," said Johnny.

268° *Method.*   A little girl was doing her arithmetic homework on the living room floor and kept calling to her mother in the kitchen for assistance. "What is 129/3?" she called. Her mother gave her the answer. "What is 196/7?" she next called, and her mother gave her the answer. So it went for a number of problems when the mother finally asked, "Why don't you work out some of the problems yourself?" "Oh," explained the little girl, "our teacher said we could use any method we wished."

269° *Addition.*   The students were sent to the board to add 18, 27, and 34, while the teacher circulated around to see how they were doing. When she came to the overgrown athlete of the class, she found him in his gym jersey standing pleased beside his work. On the board he had figured:

18
27
34
———
HIKE

270° *A difficult problem.*   H. S. Vandiver, the leading expert on Fermat's last theorem and related problems, discussed in a paper the problem of deciding whether there are infinitely many primes p for which $(p - 1)! + 1$ is divisible by p. He remarked: "This question seems to me to be of such a character that if I should come to life at any time after my death and some mathematician were to tell me that it has definitely been settled, I think I would immediately drop dead again."—LEO MOSER

QUADRANT FOUR

From Schopenhauer on arithmetic
to a matter of punctuation

COMPUTERS

M A N Y of the stories one hears about the sophisticated electronic computers tend to ridicule the machines one way or another. Perhaps this merely reflects a human awe and fear of the machines. At any rate, here are some more stories about computers.

271° *Schopenhauer on arithmetic.* Schopenhauer described arithmetic as the lowest activity of the spirit, as is shown by the fact that it can be performed by machine.—LEO MOSER

272° *E. T. Bell on electronic computers.* Eric Temple Bell was born in Scotland, in 1883, and received his early education in England. In 1902 he came to the United States, completed his undergraduate work at Stanford University, and then secured his A.M. and Ph.D. degrees at the University of Washington and Columbia University. After a succession of teaching posts at the University of Washington, the University of Chicago, and Harvard University, he was appointed, in 1926, Professor of Mathematics at the California Institute of Technology. He became a potent and influential force in American mathematics. He was a noted number theorist and had a deep interest in contemporary science. As a singularly fluent, gifted, highly knowledgeable, and, when he wished, skillfully cutting writer of semipopular books on mathematics and its history, he became widely known. In addition to his technical works, he also wrote, under the pseudonym of John Taine, several very successful science fiction novels. He died in 1960.

Following are a pair of remarks made by Dr. Bell about the modern electronic computers:

1. "What I shall say about these marvelous aids to the feeble human intelligence will be little indeed, for two reasons: I have always hated machinery, and the only machine I ever understood was a wheelbarrow, and that but imperfectly."

2. "I cannot see that the machines have dethroned the Queen. Mathematicians who would dispense entirely with brains possibly have no need of any."

273° *The guilty party.*   A while ago, in 1971, a newspaper reported that the employees of a large British firm became annoyed by a prolonged series of blunders emanating from the pay office, and they threatened to strike if the perpetrator of the blunders was not fired. Finally the firm gave in and fired the guilty party, a modern electronic computer, and peace and happiness were restored when human bookkeepers and accountants returned to the job.

274° *A comparison.*   Two mathematicians stood dwarfed before a monstrous computer that filled up a whole large wall. After a few seconds of running time, the machine disgorged a slip of paper. One of the mathematicians took the slip, studied it thoughtfully for a moment, turned to his companion and said, "Do you realize that it would take a corps of 500 skilled mathematicians, each writing at the rate of three digits per second and working day and night around the clock, over 1500 years to make a mistake of this size?"

275° *A cartoon by Dunagin.*   A young and pretty second-grade school teacher, standing behind her desk, addresses a little boy standing next to his front-row seat and holding a small battery-operated machine. The teacher says: "No, Rodney, you must tell me what two plus two is without referring to your electronic calculator."

276° *Translation of languages by a computer.*   There is a widely circulated story about an engineer who programmed a computer to translate from any language to any other language. To demonstrate his program at a technical gathering, the engineer asked for someone in the audience to suggest a phrase for the machine to translate, and obtained "Out of sight, out of mind." He entered this phrase into the computer and then asked for someone to suggest a language into which the phrase should be translated, and obtained "Russian." This instruction was put into the machine, and shortly the computer printed out a Russian phrase. But no one present understood Russian, and so the audience was perplexed as to the satisfactoriness of the translation. Finally someone came up with the bright idea that the Russian phrase should now be fed into the machine with instructions to translate it back into English. The engineer did this, and after a moment the machine printed out that "Out of sight, out of mind" first translated into Russian and then back into English resulted in the phrase "Blind idiot."

A similar experiment with "The spirit is willing, but the flesh is weak," resulted in "The whiskey's O.K., but the meat is lousy."

277° *The sophisticated computer.* Visitors are welcome at a firm that makes modern sophisticated electronic computers, and twice a day in a little auditorium a free forty-minute lecture is given during which interested visitors may hear about the marvels of the machines and see one of them put through its incredible paces. Seated at a small desk at the entrance of the auditorium is an attendant who, as visitors enter the auditorium, records on a little pad, marks that look like the following:

278° *Understanding.* At a recent exhibition of electronic computing machines in the Netherlands, Queen Juliana remarked that not only could she not understand these machines, but she could not understand the people who could understand them.

—LEO MOSER

279° *The longest root.* A real number is called *simply normal* if all ten digits occur with equal frequency in its decimal representation, and it is called *normal* if all blocks of digits of the same length occur with equal frequency. It is believed, but not known, that π, e, and $\sqrt{2}$, for example, are normal numbers. To obtain statistical evidence of the supposed normalcy of the above numbers, their decimal expansions have been carried out to great numbers of decimal places.

In 1967, British mathematicians, working with a computer, carried the decimal expansion of $\sqrt{2}$ to 100,000 places. In 1971, Jacques Dutka, of Columbia University, found $\sqrt{2}$ to over one million places—after 47.5 hours of computer time, the electronic machine ticked off the decimal expansion of $\sqrt{2}$ to at least 1,000,082 correct places, filling 200 pages of tightly spaced computer print-out, each page containing 5000 digits. This is the longest irrational root so far computed.

280° *Computerized art.*   A number of the advanced calculators allow peripheral devices to be attached to them. Since these devices are controlled by the calculator, they may be regarded as programmable. Among the devices are plotters which, using coordinates developed by the calculator under program control, can be made to draw "curves." Although only straight line segments connecting pairs of points can be drawn, by making the line segments very short, curves can be very nicely approximated.

In 1971, the Hewlett-Packard Company, a manufacturer of these latest calculator-plotter machines, sponsored an art contest for their users. The winning entries of the Calculator Art Contest were published in the Hewlett-Packard journal *Keyboard,* which is devoted to uses of the company's calculators and peripherals. The first prize was awarded to Paul Milnarich of El Paso, Texas, for his WAVES, a striking three-dimensional-looking figure of several concentric ovals formed by a succession of sinusoidal curves obtained from the equation

$$f(x,y) = [\exp(2\sqrt{(x^2 + y^2)}/1500)][\sin 2\sqrt{(x^2 + y^2)}].$$

Second prize went to G. Winston Barber of Philadelphia, Pennsylvania, for his figure EFFIGY, a weird face-like mask, wherein the equation

$$R = A_i(B + \sin n\theta_j),$$

with various limits on A and θ, was used to produce different parts of the figure. Third prize was awarded to John A. Ashbee, of

Auburn, California, for his PLAYMATE, an attractive line drawing of a seated female nude made up of 42 fitted pieces from the graph of a fourth-degree polynomial of the form

$$y = b_0 + b_1 x + b_2 x^2 + b_3 x^3 + b_4 x^4.$$

Runners-up in the contest were: SPIROGRAM, by Lt. Ronald P. Krahe of San Antonio, Texas; UNNAMED, by N. M. Baker of Greenville, Texas; RANDOM PLOT, by W. E. Shepherd of San Diego, California; ISOMETRIC REPRESENTATION OF A THREE-DIMENSIONAL OSCILLATOR, by Fred B. Otto of Orono, Maine; and INFINITY, by Peter Zimmerman of Westport, Connecticut.

281° *Computer chess.*   In his article "The robots are coming —or are they?" in the May 1976 *Chess Life & Review*, International Chess Master David Levy points out that interest in computer chess has been steadily increasing over the past two years, with a dramatic increase in chess programs being written in the United States and Canada. There have been two computer tournaments in Europe. The first World Computer Chess Championship match was held in Stockholm in 1974 and was won by KAISSA, a program written at the Institute of Control Science in Moscow. KAISSA scored 100%, finishing ahead of four entries from the United States, three from Great Britain, and one each from Austria, Canada, Hungary, and Norway.

The best-known computer chess competition is the annual tournament sponsored by A.C.M. (the Association for Computing Machinery). In October, 1975, the sixth A.C.M. competition took place in Minneapolis and was won by the program CHESS 4.4 of Northwestern University. Earlier versions of this program had won the first four A.C.M. matches, but in the fifth competition, played in San Diego in 1974, first place was taken by the Canadian program TREEFROG.

In spite of the rapid progress being made in computer chess, Levy feels confident that no program will be able to beat him in any match prior to August, 1978, and he has offered a $2500 bet to that effect. Levy feels that it will be at least 25 years before programs will play sufficiently well to earn the FIDE International Master title.

282° *Computer stamps from The Netherlands.* Computers have invaded the art world. On April 7, 1970, The Netherlands issued a set of five postage stamps containing designs made with a computer coupled to a plotter. The computer was a CORA I, at the Technological University in Eindhoven. The attached plotter was able to draw straight line segments connecting points whose coordinates were either given or calculated by part of the program. It could also draw circles or parts of circles. The program included one for rotating the coordinate axes separately through given angles, one for translation in the direction of either axis, and one for changing the scale in the direction of either axis. The CORA I can be used in combination with larger computers for drawing graphs. Two of The Netherlands stamps were made by having a part of the computing done on the EL-x8 at the Technological University.

All five stamps are 25×36 mm, and were printed in two colors in sheets of 100 by Joh. Enschedé en Zonen in Haarlem, who are also the printers of Dutch banknotes.

The 12c stamp shows an axonometric projection of a cube, the faces of which contain a central circle surrounded by a set of concentric superellipses $(x/a)^n + (y/b)^n = 1, n > 2$.

The 15c stamp employed a perspective program to draw several cubes subdivided into smaller cubes, wherein all vertical lines are omitted.

The 20c stamp contains two perpendicular line segments rotated in opposite directions through 90°.

The 25c stamp employed an affine program on circles, resulting in three double sets of nested ellipses.

The 45c stamp is made up of four spirals winding about a common origin.

The designs of the stamps were originated by R. D. E. Oxenaar, one of the graphic artists of the Post Office Department.

283° *Formula stamps from Nicaragua.*   While on the subject
of postage stamps, we might narrate the following, even though
there is no connection with electronic computers.

In 1971 Nicaragua issued a series of postage stamps paying
homage to the world's "ten most important mathematical formulas." Each stamp features a particular formula accompanied
by an appropriate illustration, and carries on its reverse side a
brief statement in Spanish concerning the importance of the
formula.

The first stamp in the series is devoted to the fundamental
counting formula "$1 + 1 = 2$," and pictures an ancient Egyptian
grasping the concept of counting. Other early mathematical
achievements honored on these stamps are the Pythagorean relation "$a^2 + b^2 = c^2$," and the Archimedean law of the lever
"$w_1 d_1 = w_2 d_2$," the one so basic in geometry and other so basic
in engineering. As stamps honoring later achievements, there is
one devoted to John Napier's invention of logarithms and one to
Sir Isaac Newton's universal law of gravitation. Among more modern formulas honored on these stamps are J. C. Maxwell's four
famous equations of electricity and magnetism, Ludwig Boltz-
mann's gas equation, Konstantin Tsiolkovskii's rocket equation,
Albert Einstein's famous mass-energy equation "$E = mc^2$," and
Louis de Broglie's revolutionary matter-wave equation.

It must be pleasing to scientists and mathematicians to see
these formulas so honored, for these formulas have certainly contributed far more to human development than did many of the
kings and generals so often featured on postage stamps.

284° *A computer triumph.*   For many years (since shortly
after 1850) one of the most celebrated unsolved problems in mathematics has been the famous conjecture that four colors suffice to
color any map on a plane or a sphere, where in the map no two
countries sharing a common linear boundary can have the same
color. An enormous amount of effort has been expended on this
problem and many partial results have been obtained, but the
problem itself remained refractory. Then, in the summer of 1976,

Kenneth Appel and Wolfgang Haken of the University of Illinois, established the conjecture by an immensely intricate computer-based analysis. The proof contains several hundred pages of complex detail and subsumes over 1000 hours of computer calculation. The method of proof involves an examination of 1936 reducible configurations, each requiring a search of up to half a million logical options to verify reducibility. This last phase of the work occupied six months and was finally completed in June, 1976. Final checking took the entire month of July, and the results were communicated to the *Bulletin of the American Mathematical Society* on July 26, 1976.

The Appel–Haken solution is unquestionably an astounding accomplishment, but a solution based on computerized analyses of close to 2000 cases with a total of 10 billion logical options is very far indeed from elegant mathematics. Certainly on at least an equal footing with a solution to a problem is the elegance of the solution itself. This is probably why, when the above result was personally presented by Haken to an audience of several hundred mathematicians at the University of Toronto in August, 1976, the presentation was rewarded with nothing more than a mildly polite applause.

285° *Computeritis.* Unfortunately, there is a developing feeling, not only among the general public but also among young students of mathematics, that from now on any mathematical problem will be resolved by a sufficiently sophisticated electronic machine, and that all mathematics of today is computer-oriented. Teachers of mathematics must combat this disease of *computeritis*, and should constantly point out that the machines are merely extraordinarily fast and efficient calculators, and are invaluable only in those problems of mathematics where extensive computing or enumeration can be utilized.

MNEMONICS

A MNEMONIC is any aid to the memory. Mathematicians have concocted a number of mnemonics for recalling certain important principles, formulas, and numbers. For example, in elementary trigonometry classes the students learn "<u>A</u>ll <u>s</u>tudents <u>t</u>ake <u>cal</u>culus" (or, in New York State, "<u>A</u>lbany <u>S</u>tate <u>T</u>eachers <u>C</u>ollege") to recall that in quadrant one <u>a</u>ll the trigonometric functions are positive, in quadrant two only the <u>s</u>ine and its reciprocal, in quadrant three only the <u>t</u>angent and its reciprocal, and in quadrant four only the <u>c</u>osine and its reciprocal.

We have already, in Items 41°, 119°, and 145° of *Mathematical Circles Revisited,* given examples of mnemonics in mathematics. Here are some more.

286° *A French mnemonic for pi.* By replacing each word by the number of letters it contains, the following French poem yields π correct to 26 decimal places.

> Que j'aime à faire apprendre
> Un nombre utile aux sages
> Immortel Archimède artiste ingénieur
> Qui de ton jugement peut priser la valeur
> Pour moi ton problème
> A les pareils avantages!

The last line leads to the sequence of digits 1379; for the decimal expansion of π it should be 3279.

M. H. Greenblatt says that George Gamov once wrote an article for *Scientific American* (Oct. 1955), in which he gave the first five decimals in the expansion of π as 3.14158. Later, a reader wrote to *Scientific American,* chiding Gamov for being wrong in the fifth decimal place. Gamov, in his apology, explained that the error was due to the fact that he was an atrocious speller; he remembered π from the above French poem, but he had spelled the word "apprendre" with only one "p."

287° *Another mnemonic for pi.*

> Sir, I bear a rhyme excelling
> In mystic force and magic spelling
> Celestial sprites elucidate
> All my own striving can't relate.
> 3.14159 / 265358 / 979 / 323846

288° *An impossible mnemonic.*   In Item 41° of *Mathematical Circles Revisited,* four sentence mnemonics are given for recalling the decimal expansion of π to a number of decimal places. For example, if in the sentence

> May I have a large container of coffee?

one should replace each word by the number of letters it contains, one would obtain π expressed to 7 decimal places (3.1415926). The most successful mnemonic in the above reference gives π to 30 decimal places. No one has ever been able to make up a mnemonic of this kind giving π to more than 31 decimal places. Why is this?

289° *Rational mnemonics.*   The number π can be approximated by rational numbers. For example:

$$22/7 = 3.14|28,$$
$$355/113 = 3.141592|92,$$
$$104348/33215 = 3.141592653|92142,$$
$$833719/265381 = 3.14159265358|108,$$

which, in turn, give π correct to 2, 6, 9, and 11 decimal places. In the July 1939 issue of *The Mathematical Gazette,* the following mnemonics were given for recalling the last two fractions:

$$\frac{\text{calculator will get fair accuracy}}{\text{but not to } \pi \text{ exact}} \quad,$$

$$\frac{\text{dividing top lot through (a nightmare)}}{\text{by number below, you approach } \pi} \quad.$$

290° *A mnemonic in trigonometry.* Every trigonometric identity remains valid when each trigonometric function is replaced by the corresponding hyperbolic function provided we change the SIGN of each term that contains a product of *two* SINES.

—GEORGE POLYA and GORDON LATTA
Complex Variables
John Wiley & Sons, Inc., 1974, p. 69

291° *Recalling the Poisson distribution.* The Poisson distribution states that if m happenings occur on the average, then the probability that n will occur is

$$P_m(n) = (m^n e^{-m})/n!.$$

M. H. Greenblatt has noted the following remarkable mnemonic, involving, in order, the first six letters of the word *mnemonic* itself, for recalling this formula:

(m to the n, e to the minus m, over n factorial).

292° *Easy.* A mathematician had trouble remembering the local call number, 2592, of his home telephone number. Finally, after several hours of concentrated work, he announced, "I have it. The number 2592 is the unique solution in distinct positive integers x, y, z of the expression $xyzx = x^y z^x$, where the left side represents a decimal expression and the right side is a product of powers."

M. H. Greenblatt has reported this as a true story. The phone number concerned was that of B. Rothlein, who, between 1943 and 1947, lived on 41st Street in Philadelphia.

THE NUMBER THIRTEEN

293° *The one-dollar bill and the number thirteen.* How many people know the following about the United States one-dollar bill?

1. The incomplete pyramid on the back has 13 steps.
2. Above the pyramid appear the words "Annuit Coeptis," which contain 13 letters.

3. The American bald eagle holds in one talon an olive branch with 13 leaves, in the other talon a bundle of 13 arrows.

294° *Richard Wagner and the number thirteen.* Richard Wagner's name contains 13 letters. He was born in 1813 and $1 + 8 + 1 + 3 = 13$. He composed 13 great works of music. *Tannhäuser,* one of his greatest works, was completed on April 13, 1845, and it was first performed on March 13, 1861. He finished *Parsifal* on January 13, 1842. *Die Walküre* was first performed in 1870 on June 26, and 26 is twice 13. *Lohengrin* was composed in 1848, but Wagner did not hear it played until 1861, exactly 13 years later. He died on February 13, 1883; the first and last digits of this year form 13.

295° *A prosaic explanation.* Claude Terrail, the proprietor of the luxurious La Tour d'Argent Restaurant, overlooking the Seine in Paris, explains the superstition about having thirteen people seated at a table. "The reason is," he says, "that most people have sets of only twelve knives, forks, and dinner plates."

MERSENNE NUMBERS

296° *Father Mersenne and his numbers.* Numbers of the form $M_n = 2^n - 1$, where $n = 1, 2, 3, \ldots$, are called Mersenne numbers, after Father Marin Mersenne (1588–1648), a Minimite friar who taught philosophy and theology at Nevers and Paris and who maintained an assiduous correspondence in an almost indecipherable handwriting with the top mathematicians of his day. Although Mersenne was a voluminous writer, it was in the theory of numbers, particularly in connection with prime and perfect numbers, that he left his most lasting mark.

Mersenne numbers are interesting for two reasons—the largest known prime numbers are Mersenne numbers, and it is with the help of prime Mersenne numbers that we discover perfect numbers (numbers which are equal to the sum of all their natural divisors that are less than the numbers themselves).

In a statement in the preface of his *Cogitata physico-mathematica,* published in 1644, Mersenne implied that the only values of n not greater than 257 for which M_n is prime are

$$2, 3, 5, 7, 13, 17, 19, 31, 67, 127, \quad \text{and} \quad 257.$$

This statement of Mersenne's has proved to be incorrect, both by omission and commission. For it has been shown that for $n = 61$, 89, and 107, M_n is prime, and for $n = 67$ and 257, M_n is composite. The complete list of currently known values of n for which M_n is prime is

$$2, 3, 5, 7, 13, 19, 31, 61, 89, 107, 127, 521, 607, 1279,$$
$$2203, 2281, 3217, 4253, 4423, 9689, 9941, 11212, \quad \text{and}$$
$$19937.$$

It took modern electronic computers to establish the primality of the very large numbers M_n for those values of n in the list exceeding 127; the last and largest one was established in 1971.

The primality of the 19-digit number M_{61}, the first case overlooked by Mersenne, was established by P. Seelhoff in 1886 and J. Pervušin in 1887. The composite nature of M_{67}, the first incorrect case included by Mersenne, was established by E. Fauquembergue in 1894 (by a process not yielding any factors of the number) and by F. N. Cole in 1903 (by expressing M_{67} as a product of two large prime numbers).

For more than 75 years, until 1952, the 39-digit number M_{127} was the largest known prime number.

297° *A well-received paper.* E. T. Bell, in his engaging book *Mathematics—Queen and Servant of the Sciences,* relates an amusing story in connection with the paper given by F. N. Cole (1861–1927) in 1903 on the factorization of the Mersenne number $M_{67} = 2^{67} - 1$.

At the October, 1903, meeting in New York of the American Mathematical Society, Cole had a paper on the program with the modest title *On the factorization of large numbers.* When the chairman called on him for his paper, Cole—who was always a man of very few words—walked to the board and, saying nothing, proceeded to chalk up the arithmetic for raising 2 to the sixty-seventh power. Then he carefully subtracted 1. Without a word he moved over to a clear space on the board and multiplied out, by longhand,

$$193,707,721 \times 761,838,257,287.$$

The two calculations agreed. . . . For the first and only time on record, an audience of the American Mathematical Society vigorously applauded the author of a paper delivered before it. Cole took his place without having uttered a word. Nobody asked him a question.

Some years later, in 1911, Bell asked Cole how long it had taken him to crack M_{67}. Cole replied: "Three years of Sundays."

298° *A pastime.*   To while away time during the occupation in the Second World War, the French mathematician Paul Poulet (d. 1946) worked out the prime decomposition of the large Mersenne number $M_{135} = 2^{135} - 1$, finding

$$M_{135} = (7)\ (31)\ (73)\ (151)\ (271)\ (631)\ (23,311)\ (262,657)$$
$$(348,031)\ (49,971,617,830,801).$$

BUSINESS MATHEMATICS

299° *Obtaining a receipt.*   "I loaned a tricky competitor $1000," sighed a businessman, "and he has not returned me a receipt. What can I do?"

"Write sternly," advised his friend, "and demand immediate payment of the $2000."

"You did not hear me correctly," replied the businessman. "I said the loan was $1000."

"I know," nodded the friend, "and your competitor will indignantly write you and tell you so. Then you will have your receipt."

300° *Honest business.*   An angry customer returned to a jewelry store and demanded a refund on a watch that he had bought there a few days earlier. "This watch," he fumed, "loses fifteen minutes every hour."

"Of course it does," agreed the jeweler. "Didn't you see the sign '25 percent off' when you purchased it?"

301° *No accounting.* A recent newspaper ad of the Oklahoma School of Accountancy was headed: "Short course in Accountancy for women." Not long after the ad appeared, a note reached the school's president. It read: "There is no accounting for women."—*Tulsa Tribune*

302° *Mathematics of finance.* A high-school math teacher received a form letter from a loan company stating, "Because you are a teacher you can borrow $100 to $1000 from us simply by mail." The teacher's reply said, "Perhaps I can borrow from you because I am a teacher, but I would not be able to pay it back to you for the same reason."

303° *Cinderella's pumpkin.* One evening a professor from the Harvard Business School told his six-year-old son the story of Cinderella. The youngster paid close attention, particularly when the father came to the part of the story where the pumpkin turns into a golden coach. "Say, Dad," he interrupted at that point, "Did Cinderella have to report that as straight income, or was she permitted to call it a capital gain?"

304° *Depreciated currency.* When I was a kid, ten cents was a lot of money. My, how dimes have changed.

305° *A visit to a scientific shrine.* Two American physicists touring Italy visited the famous Leaning Tower of Pisa, since it was here that Galileo dramatically demonstrated that, contrary to the teachings of Aristotle, bodies of different weights dropped from the same height reach the ground in the same time. While parking their car, a uniformed attendant handed the scientists a pink ticket and collected 100 lire. When the scientists later returned to their hotel they inquired of the concierge, "Who gets the money collected for parking at the Leaning Tower?"

The concierge looked puzzled, examined the ticket, smiled,

and exclaimed, "There's no parking charge anywhere in Pisa. What you did was to insure your car against damage should the Leaning Tower fall on it."

PROBABILITY AND STATISTICS

306° *An easily understood picture.*   Suppose the nearly three billion persons on the earth were compressed into a single town of 1000 inhabitants. Then:

1. 303 persons would be white, 697 non-white.

2. 60 persons would represent the U.S.A., 940 all the others.

3. The 60 Americans would receive one-half the town's income, 940 the other half.

4. The 60 Americans would have a life expectancy of over 70 years, the other 940 a life expectancy of under 40 years.

5. The 60 Americans would consume 15 percent of the town's food supply, and the lowest income group of the Americans would be better off than the average of the 940.

6. The 60 Americans would use 12 times as much electricity, 22 times as much coal, 21 times as much oil, 50 times as much steel, and 50 times as much equipment as all 940 remaining members of the town.

307° *Hydrodamnics.*   There seems to be a newly emerging branch of probability that may be called *hydrodamnics,* concerned with the persistent cussedness of inanimate objects. It is hoped that hydrodamnics will finally resolve the reasons why:

1. a piece of buttered bread always falls to the floor with the buttered side down;

2. juice squirting from a grapefruit always goes directly into one's eyes;

3. a wind always blows out your last match when lighting a campfire.

Any reader can easily continue the list.

308° *Statisticians.* It was Mark Twain who divided all prevaricators into three classes: liars, damn liars, and statisticians. Someone else described a statistician as a person who, with his head lying in a heated oven and his feet packed in ice, says, "On the average, I feel fine."

309° *An exercise in probability.* A wise man was visited by a large delegation of malcontents who poured out their troubles to him. The wise man said, "Each of you write down your greatest trouble on a piece of paper." He then collected all the papers and dropped them into a large pot. "Now each of you draw a paper from the pot. By the laws of probability, essentially all of you will have a brand-new trouble to worry over."

The malcontents followed the wise man's suggestion. When they read the new troubles, each one begged to have his own trouble back.

310° *The odds.* A Las Vegas visitor awoke in his hotel one night with severe pains in the stomach, and put in an emergency call for the house physician. After a quick examination, that gentleman folded his stethoscope and said, "I'll give you four to one you have acute appendicitis."

311° *A safe risk.* An upset insurance inspector was scolding a new agent. "Why in the world did you write a policy on a man 98 years old?" he demanded. "Well," replied the agent, "I consulted the census report and found that there were only a very few people of that age who died each year."

312° *Statistics.* It has been said that one in every four Americans is unbalanced. Just think of your three closest friends. If they seem all right, then you're the one.

313° *A random choice.* A fellow insists his name is Seven-and-One-Eighth Flannery. He explains that his parents picked his name out of a hat.

314° *Too risky.*   "I'm truly sorry, Max," said the probability instructor, "but if I excuse you from class today, I'd have to do the same for every other married student in the class whose wife gave birth to quintuplets."

315° *A statistician protects himself.*   A certain statistician traveled considerably about the country giving popular lectures in his field of interest. His traveling, of course, had to be done by plane, a mode of journeying that caused the man great uneasiness, especially in view of the recent cases of bombs hidden aboard planes. To help alleviate his uneasiness, he calculated the probability of traveling on a plane carrying a bomb, and was pleased to find the probability very low. He then calculated the probability of traveling on a plane carrying two bombs, and was highly relieved to find this probability to be infinitesimal. So, thenceforth, in his travels about the country, the statistician always carried a bomb with him.

ALGEBRA

316° *Positive-negative.*   A usually soft-hearted father took a firm stand against one of his 17-year-old daughter's way-out demands. Sensing the finality of the "No" she was receiving, the daughter gave up instead of trying to wear her father down. "What's the use?" she sulked, "Daddy's in one of his positive-negative moods."

317° *A query.*   A little boy once asked, "Why is it that so many churches have plus signs on them?"

318° *A "proof" that "$(-)(-) = +$."*   I obtained the following elegant "proof" that "$(-)(-) = +$" from Raphael Mwajombe of Tanzania:
Imagine a town where good people are moving in and out and bad people are also moving in and out. Obviously, a good person is $+$ and a bad person is $-$. Equally obvious, moving in is $+$ and

moving out is −. Still further, it is evident that a good person moving into town is a + for the town; a good person leaving town is a −; a bad person moving into town is a −; and, finally, a bad person leaving town is a +. Our results are neatly summarized in the following table:

	moving in (+)	moving out (−)
good person (+)	+	−
bad person (−)	−	+

—Roy Dubisch
The Mathematics Teacher, Vol. LXIV, No. 8 (Dec. 1971), p. 750

319° *Some nonassociative phrases.*

1. He is in the [high (school building)].
   He is in the [(high-school) building].

2. They went into the [dark (green house)].
   They went into the [(dark green) house].

3. They don't know how [good (meat tastes)].
   They don't know how [(good meat) tastes].

320° *Much ado about nothing.* Professor Morris Marden, who in 1975 retired from the University of Wisconsin at Milwaukee, spent over forty years studying the *zeros* of functions. As a result he has long been teased as the world's expert on *nothing*.

321° *The coconut problem.* The following is a well-known problem: "Five sailors, *A, B, C, D, E*, gathered a great pile of coconuts, which they agreed to divide equally the next morning. They had a pet monkey. Sailor *A* thought he would make sure of his share, and got up secretly in the night, divided the nuts into five equal piles, and finding an extra one left, gave it to the monkey.

Then, concealing his pile, he heaped the other four piles into one and went back to sleep. Shortly afterwards, *B* woke and did just what *A* had done: he made five equal piles, giving an odd coconut to the monkey, concealed his own pile, heaped the other four piles together, and went back to sleep. Next *C* awoke and did the same; then *D* did likewise; and finally so did *E*. In the morning they did as they had planned, and this time the nuts came out even in five equal shares with no extra one for the monkey. Find the (minimum) number of coconuts originally gathered."

Interest in this problem received great popular stimulation in 1926, for in that year, in the October 9th issue of *The Saturday Evening Post*, the entertaining writer Ben Ames Williams used the problem in a story. In the story the problem was employed to distract a man who should have been calculating on a contract, and it prevented the man from getting an important bid in on time.

Answer: The minimum number is 3121. For four sailors the minimum number would have been 765; for three sailors, 25; for two sailors, 11.—WILLIAM R. RANSOM

322° *A clever solution.*   Enrico Fermi (1901–1954), who designed the first atomic piles and in 1942 produced the first nuclear chain reaction, has been credited with the following singularly clever solution of the coconut problem considered in Item 321°.

The number -4 is clearly a mathematical solution, because the first sailor, finding -4 coconuts, gives 1 to the monkey, leaving -5, then takes his fifth, leaving -4 again for the next sailor, and so on. To get the smallest positive solution, we must add the smallest positive number divisible by 5 five times, which is 5^5. Thus the answer to the problem is

$$5^5 - 4 = 3125 - 4 = 3121.$$

323° *A coach's formula.*   Mr. Webb, a very successful coach at St. John's, Cambridge, was a good teacher and an amusing personality. He used to say, "$pp > jj$" (plodding patience is greater than jumping genius). He could be very sarcastic, and would say to a student, "Write all you know on this piece of paper"—something about an inch square.—L. J. MORDELL

324° *The metric system.*   Part of the work in present-day algebra classes is converting from English measurement to metric measurement, and vice versa.

One of Dunagin's masterful cartoons (it appeared in the Friday, September 12, 1975 issue of the *Bangor Daily News*) shows a man enthusiastically addressing members of the U.S. Government Dept. of Commerce. With charts and conversion plans at hand, he is stressing the importance of the coming change-over from the English to the metric system of measurements. With a warning finger in the air, he says: "I want everyone to think metric! Remember an ounce of prevention is worth a pound of cure!"

325° *Pointing the way.*   The nation's first metric-system highway signs were erected along Interstate 71 in Ohio. Thus a sign outside Columbus now reads: CLEVELAND—94 MILES—151 KILOMETERS. The new signs are intended to prepare motorists for the nationwide conversion, which is expected to occur by 1983.

—*Tomorrow's World*

326° *Andre Weil's exponential law of academic decay.*   "First-rate people appoint first-rate people; second-rate people appoint fourth-rate people, who appoint eighth-rate people, and so on."

GEOMETRY

327° *The fourth dimension in landscape planning.*   In designing a landscape plan for a piece of property, the width and depth of the property, the position of the buildings on the property, the rise and fall of the land, and the height of the buildings, of neighboring structures, and of the existing large trees, collectively constitute three of the dimensions that must be included in the plan. But there is a fourth dimension, the passage of time, that must also be included, and this dimension is one of the most important considerations in the design, and is a dimension that does not show up in the initial planning process.

Many a landscape design has been ruined by a failure to take care of the fourth dimension. Trees and shrubs, that in time grow very large, are planted too close together or too close to view-commanding windows. Shrubs that reach only a moderate height are later hidden by other taller-growing items. It takes real skill and vision on the part of the landscape architect to prepare a design that, when executed, will look attractive, not only on completion of grading, growing of grass, and planting of trees and shrubs from the nursery, but will continue so as time marches on. Indeed, the truly good plan will be attractive at the start and then actually gain in attractiveness with the passage of the fourth dimension. The landscape architect must constantly ask himself how his developing design will look five years hence, ten years hence, twenty years hence.

328° *Applicable to angle trisectors and circle squarers.*   William G. McAdoo, Secretary of the Treasury in President Woodrow Wilson's Cabinet from 1913 to 1918, claimed: "It is impossible to defeat an ignorant man in argument," a statement certainly applicable to angle trisectors and circle squarers. Apropos these latter people, Garrett Hardin, in his excellent book *Nature and Man's Fate*, says: "Those who have had any contact with angle trisectors, circle squarers, or the inventors of perpetual-motion machines will attest to their being a very queer bunch of people, indeed. It is not their proposals that merit study, but their personalities. What defect in their character is it that makes them unwilling to accept the idea that perhaps they cannot have everything they want? Whatever it is, it is akin to the immaturity of the spoiled child and the compulsive gambler."

329° *A strong theorem.*   In a paper on the foundations of Euclidean plane geometry, R. H. Bruck explains:

"The axioms of incidence require so little of the Euclidean plane that very few theorems have been proved. Indeed, the main theorem—I call it Hall's theorem—might be stated as follows: Any damn thing can happen."—LEO MOSER

330° *To meet or not to meet.*   There are geometries in which parallel lines do not exist. Of such a study one may say: "The reason parallel lines never meet in this geometry is because they were never introduced."

331° *Careful use of terms.*   A geometry professor was induced to introduce the daughter of one of his colleagues to his son. To prepare the son, the professor first described the young lady to the lad. After it all was over, the boy, who hadn't been too impressed with the girl, said to his father, "I thought you said the young lady's legs were without equal." "Not at all," replied the professor, "I said they were without parallel."

332° *Victor Borge's metric.*   "Laughter is the shortest distance between two people."

333° *Martian geometry.*   Harry Schor, of Far Rockaway High School on Long Island, says that he is teaching geometry to a class of such peculiar students that he has to prove right triangles congruent by "hypotenuse and tentacle."—ALAN WAYNE

334° *Volume.*   Mrs. Jones was busily applying light yellow paint to the dark brown wall-board inside her garage. Neighbor Smith commented that the lighter color would make the garage seem larger. "I certainly hope so," said Mrs. Jones, "We could really use the extra space."

335° *Circles.*   A first-grade teacher asked her pupils to draw pictures showing what their fathers did for a living. She noticed one child drawing circles on his paper, and asked, "What does your father do for a living?"

The child replied, "He's a doctor, Mrs. Sawyer. He makes rounds."

336° *A possible misunderstanding.* In the "continuous geometry" developed by von Neumann, points, as such, play no role.
Originally von Neumann proposed another name for his geometry, but his colleagues pointed out that this name might lead to
misunderstanding. Von Neumann's proposed name was "point-
less geometry."—LEO MOSER

337° *An oblate spheroid.* There was a politician who can be
said to have resembled the earth. The politician, you see, was
severely clobbered in a bid for re-election to Congress, and the
earth is an oblate spheroid—that is, it is flattened at the poles.

338° *Angles.* An Eskimo father sat in his igloo reading nursery rhymes to his little boy. "Little Jack Horner sat in a corner,"
began the father, when the boy interrupted with, "Daddy, what is
a corner?"

RECREATIONAL MATTERS

339° *Loculus Archimedes.* One of the earliest examples of a
"cut-out" puzzle that has come down to us is that which has become known as the *loculus Archimedes,* or the *box of Archimedes.* It
consists of assembling the fourteen pieces (11 triangles, 2 quad-
rilaterals, 1 pentagon) pictured in Figure 5, top, into a square. It
is a teasing puzzle that makes a nice gift when the pieces are neatly
cut from thin wood or plastic and are to be assembled to fill the
bottom of an accompanying shallow square box, as in Figure 5,
bottom.

Historians feel that we cannot ascribe the puzzle to Archimedes, and explain the attachment of his name to it as probably only
a way of asserting its cleverness. Indeed, the phrase "Archimedean
problem" has come proverbially to mean anything that is difficult.
Since we are ignorant of the origin of the puzzle, we do not know
why this particular dissection of a square was chosen.

340° *Storage place for imaginary numbers.* During the 1976
school year, I received, through the mail, a little square box of

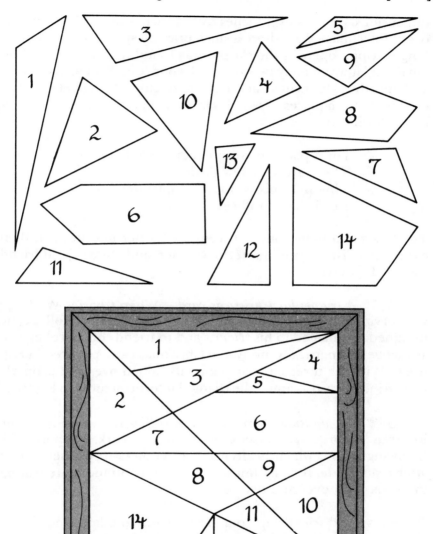

FIGURE 5

139

dimensions 4 inches by 4 inches by 2¼ inches deep. The box was beautifully finished and bore a small brass clasp and a pair of brass hinges. Upon opening the clasp it was found that the box had no "insides;" it was merely two pieces of wood, one ¾ of an inch thick and the other 1½ inches thick, hinged together. Lying between the top and bottom pieces of wood was a small piece of paper bearing the following rhyme:

> For a place to store imaginary numbers,
> There are very few devices.
> Although this is not a complex box,
> I feel that it suffices.

The box was a humorous gift from my former fine student, then excellent assistant, now masterful teacher, and always good friend Carroll F. Merrill.

341° *A triangular Christmas card.*   When Charles W. Trigg was serving as Dean of Instruction at Los Angeles City College, he designed and mailed to his friends and to friends of his college an attractive geometrically motivated Christmas card, pictured in Figure 6. With the three circular segments folded over and properly tucked in, the card assumes the triangular form pictured in Figure 7.

342° *An unusual Christmas card.*   Several years ago, the Reverend George W. Walker of Buffalo, New York, composed the following poem, which is circular and endless in the sense that after reading the first six lines, you are to continue by repeating these six lines over and over endlessly.

True love and friendship come as gifts of God, and we can say
The precious gifts of God endure forever, like the way
You start to read this anywhere and presently you find
A ring like this will never end, which ought to bring to mind
We start afresh on New Year's Day, but, after that, we see
The years go on eternally, and so, it seems to me,

·   ·   ·   ·   ·   ·   ·   ·   ·   ·   ·

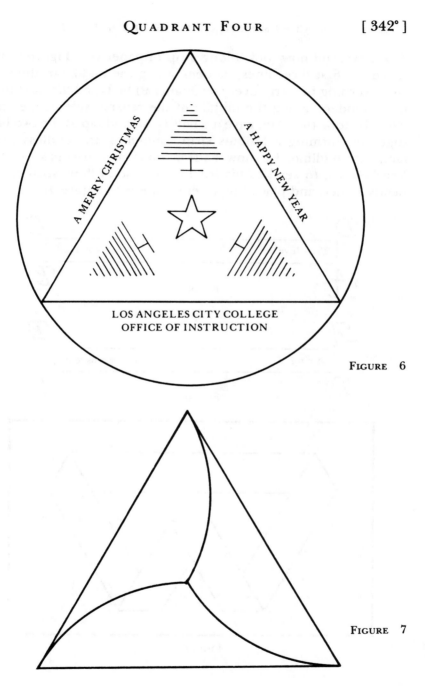

FIGURE 6

FIGURE 7

The Reverend now took a long strip of paper (see Figure 8) and typed the first three lines, tandem, along the middle of the strip. Next, turning the strip over (see Figure 9) he typed the next three lines, tandem, along the middle of the reverse side of the strip. Then he took the strip, gave it a half-twist, and taped the two ends together, forming a Möbius strip containing the endless poem along its midline. He now had an unusual Christmas, or New Year's, card, to send to his friends, for the Möbius strip can be neatly folded and placed in an envelope (see Figure 10).

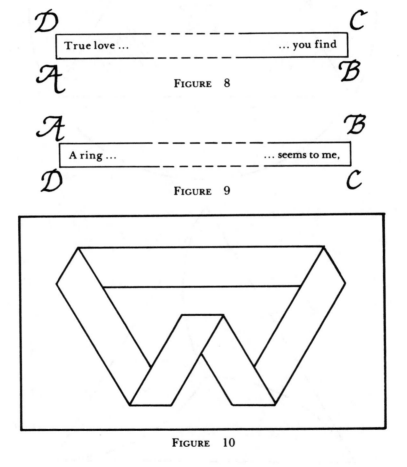

D C

True love ...   ... you find

A B

FIGURE   8

A B

A ring ...   ... seems to me,

D C

FIGURE   9

FIGURE   10

A Möbius strip of this sort can be made containing any circular and endless poem or piece of prose, such as the familiar circular and endless story:

> The night was dark and stormy, and we were all seated around the campfire smoking, when someone said, "Captain, tell us a story." So the Captain began:

343° *Another mathematical Christmas card.* Some years ago, Professor Richard V. Andree (of the University of Oklahoma, in Norman, Oklahoma) and his wife Josephine P. Andree mailed to their friends what has proved to be perhaps the best of the Christmas cards based upon the idea of graphing some equation or equations. Figure 11 shows the appearance of the front face of the card. The design on the card was actually in two colors—red for the coordinate axes and blue for everything else.

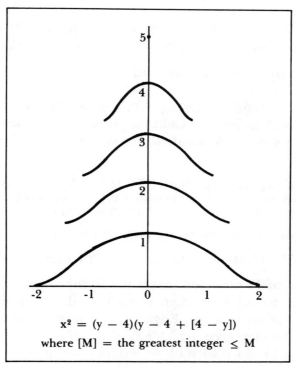

$$x^2 = (y - 4)(y - 4 + [4 - y])$$
where [M] = the greatest integer ≤ M

FIGURE 11

344° *Some season's greetings for the mathematics student.*
Many mathematical expressions have been concocted which, when
expanded or operated on in some way, yield familiar season's
greetings. For example, Professor Richard V. Andree designed the
following:

$$\begin{vmatrix} M & R^2 \\ -Y & \ln^{-1}1 \end{vmatrix} + XW,$$

where it must be realized that, using customary symbols of physics,

$$W = F \cdot S \quad \text{and} \quad F = MA.$$

Many years ago, when I was teaching in West Virginia, one of
my calculus students, then Miss Bertha Weaver, devised the following:
Given

$$y = (x + M/2)^2 - (E - x)^2/2 + (9y - 6)/3 + 2Rx + \int y \, dx$$
$$+ (14 - 6y)/2 + x^2M/2 - (2A - x)^2/4 + x(S - x/4).$$

Find dy/dx by implicit differentiation.

345° *A holiday greeting.* Professor Charles W. Trigg,
when at Los Angeles City College, proposed the following problem (*The American Mathematical Monthly*, Problem E 1241, Dec.
1956):

> The holiday greeting, *MERRY XMAS TO ALL,* is a cryptarithm in
> which each of the letters is the unique representation of a digit, and
> each word is a square integer. Find all solutions.

It turns out that there are only two solutions,

27556 3249 81 400   and   34225 7396 81 900.

Azriel Rosenfeld observed that if the further requirement is imposed that the *sum of the digits* of each word be a perfect square,
then the solution is unique. Edgar Karst concocted the allied problem: The holiday greeting, *MERRY + XMAS = TOALL,* is a
cryptarithm in which each of the letters is the unique representa-

tion of a digit and each word is divisible by 3. There is now a unique solution, 84771 + 5862 = 90633.

346° *Anagrams.*   In the December 1952 issue of *The American Mathematical Monthly* appeared the "fun" problem E 1041:

> Surely the days of anagrams are not dead. Here are four well-known mathematicians: (1) A DONUT SHIP,   (2) SHE IS A NUT,   (3) SEWER STAIRS,   (4) FIRE ON SUB.

The answers, which appeared in the June–July 1953 issue of the journal, are: (1) DIOPHANTUS, (2) STEINHAUS, (3) WEIERSTRASS, (4) FROBENIUS. A good deal of ingenuity was expressed by various solvers of the problem. For example, Charles W. Trigg produced, among others, the following alternative anagrams for the above four mathematicians: (1) THUDS PIANO, I PUT SHAD ON; (2) USES A HINT, TIN HAS USE, SHUNS A TIE, THE USA SIN; (3) WATERS RISE, SWEARS RITES; (4) I BURN FOES, FIE ON BURS, O FINE RUBS.

Trigg, Paul L. Chessin, and H. W. Gould added anagrams of further mathematicians, such as: WENT ON (NEWTON), ROT-TED HUN (TODHUNTER), REC'D ROE (RECORDE), IV TENS (STEVIN), KIND DEED (DEDEKIND), SEE GUARDS (DE-SARGUES), PELT ONCE (PONCELET), LIE CRUEL APE (PEAUCELLIER), BAN ON RICH (BRIANCHON), A FREE BUCH (FEUERBACH), RICH CASE (SACCHERI), RAM NINE (RIEMANN), SING A TUNE, MOOR (REGIOMONTANUS), I REMOVED (DE MOIVRE), MAR CONE (CREMONA), DEER GLEN (LEGENDRE), CALL APE (LAPLACE), I RIVAL ACE (CAVALIERI), SET CEDARS (DESCARTES), TIRE CHILD (DI-RICHLET), HARM A DAD (HADAMARD), I CHARM ALEC (CHARMICHAEL), A CURT NO (COURANT), LEAN MUSE (MENELAUS), REACH DIMES (ARCHIMEDES), ONE LIME (LEMOINE), NO GEM (MONGE), OR AN EPIC (POINCARE'), AS SHE (HASSE), HERE TIM (HERMITE), REGAL NAG (LA-GRANGE), NICE CORPUS (COPERNICUS), LAMED BERT (D'ALEMBERT),   A   CRITIC   (RICCATI),   ON   ETHER (NOETHER), A PINER (NAPIER), RECIPE (PEIRCE), ABLE

(ABEL), BRED IN US (BURNSIDE), I LAG SO (GALOIS), READ CHIMES (ARCHIMEDES), AM SURE NOT GO IN (REGI-OMONTANUS), CABIN FOCI (FIBONACCI), RUN A CLAIM (MACLAURIN), and several others.

On July 20, 1953, Henriette von Boeckmann, Secretary of The National Puzzlers' League, Inc. (a group that was organized on July 4, 1883), sent the following letter to the Editor of the Problem Department of *The American Mathematical Monthly:*

Dear Dr. Eves:

The National Puzzlers' League met in convention in the Hotel Statler, N.Y. City, July 3–4–5, 1953. At the business meeting one of the members, who is a subscriber to *The American Mathematical Monthly,* brought to the attention of the meeting, that your June–July issue gave some space, in their Problem Department, to a puzzle form—the Anagram.

The names of four famous mathematicians were "anagrammed," but, according to the standards of the N.P.L., they are merely muta-tions. It is true that the dictionaries define an Anagram as a rear-rangement in the letters of a world or phrase to produce some other word or phrase, but the N.P.L. narrows that meaning so that the anagrammed word must bear a relation in meaning to the original. Here are a few specimens of Anagrams as sponsored by the N.P.L.:

One smart hat—Tam O'Shanter,
So I sit and sip—Dissipations,
It's slam bang—Lambastings,
Is "anti" in "spare child"—The disciplinarians.

The N.P.L. also recognizes the Antigram, in which words express opposite meanings.

You can see that anagramming a proper name is exceedingly difficult; the result is usually a mutation.

This letter is in no sense a criticism; it is merely an attempt to keep the mental sport of puzzling on a high plane. The convention, through its President, requested its Secretary to explain to you the N.P.L.'s conception of the Anagram, which by many is considered the highest form of the puzzling art.

Yours very truly,
Henriette von Boeckmann, Secretary N.P.L.

From the point of view of The National Puzzlers' League, perhaps only WENT ON (NEWTON), KIND DEED (DEDEKIND), and ABLE (ABEL) can rate as *true* anagrams, and NO GEM

(MONGE) and I LAG SO (GALOIS) might be considered as anti-grams. It is interesting to observe that the first of our *true* anagrams was noted by Augustus De Morgan (see Item 198° of *In Mathematical Circles*).

Finally, Trigg gave what may be the only palindromic anagram of a mathematician's name: AVEC (CEVA).

347° *The great "show-me" craze.*   In the mid-1970s the great "show-me" craze erupted, like an epidemic, among the mathematically literate of the United States. The disease gained rapidly and engulfed an enormous number of the mathematical community. It raged rampantly for some time, and then subsided almost as quickly as it had arisen. During its height, its taint was to be found in a number of the country's mathematics journals. Among the victims most deeply smitten was Charles W. Trigg, as was to be expected, for the disease seemed to ferret out the keenest and most agile minds of the profession. While under the frenetic fever of the derangement, Trigg produced a prodigious number of "show-me's," of which the following can be considered only as a mere sample.

Show me a useless exponent and I'll show you one.

Show me some protesters with locked arms on a hotel floor and I'll show you a connected set.

Show me a man who counts on his fingers and I'll show you a digital computer.

Show me how to partition six and I'll show you how to do it to two, too.

Show me a glib salesman and I'll show you a line.

Show me an erratic teenager with no personal income and I'll show you a dependent variable.

Show me a community with approximately equal populations of Nordics, Africans, Chinese, and Indians and I'll show you a four-color problem.

Show me an upended maple tree and I'll show you some complex roots.

Show me a singing birthday party and I'll show you a harmonic function.

> Show me a youngster who has lost a trio of front teeth and I'll
> show you a three-space.

Another afflicted member of the mathematical fraternity was
Fred Pence, who produced the following "show-me's."

> Show me two cars enmeshed after an auto accident and I'll show
> you a rectangle (wrecked tangle).
> Show me Joe Frazier and Muhammad Ali and I'll show you ex-
> opponents.
> Show me the squares of 6, 5, and 6 and I'll show you some
> interesting figures.
> Show me a mermaid dressed in seaweed and I'll show you an
> algae bra.

Other similar viruses have at times spread through the mathematical world—who can forget, for example, the "Tom Swiftie"
craze, traces of which are still to be found in the mathematical
hinterlands. Indeed, an enterprising individual could easily assemble an interesting booklet of case histories of these mathematical
pestilences. There is no telling if some of these will, like the Asian
flu, reappear in the mathematical community, nor can we guess
what new virulent variations may afflict us in the future.

348° *Benjamin Franklin's semimagic square.*   In *Mathematical
Circles Squared* we devoted a section to magic squares; we now
consider a few more of these delightful objects, and refer the
uninitiated reader to Item 30° of the above reference for the meanings of commonly used technical terms.

Benjamin Franklin (1706–1790) interested himself in the construction of semimagic squares possessing the "bent diagonal"
property (see Item 317° of *In Mathematical Circles*). Figure 12A
shows an 8 × 8 Franklin semimagic square—each column and
each row sums to 260 (unfortunately, the two principal diagonals
fail to do so, and thus the square is only semimagic rather than
magic). But the five bent diagonals shown in Figure 12B also each
sum to 260, and there are five such bent diagonals based on the
top, five based on the right side, and five based on the left side of
the square, each summing to 260. In addition, the eight squares

52	61	4	13	20	29	36	45
14	3	62	51	46	35	30	19
53	60	5	12	21	28	37	44
11	6	59	54	43	38	27	22
55	58	7	10	23	26	39	42
9	8	57	56	41	40	25	24
50	63	2	15	18	31	34	47
16	1	64	49	48	33	32	17

A

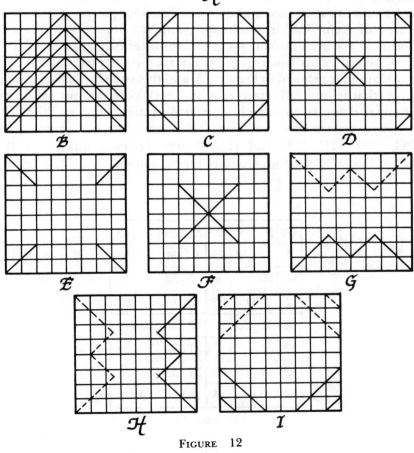

B *C* *D*

E *F* *G*

H *I*

FIGURE 12

149

indicated in Figure 12C, the eight indicated in Figure 12D, the eight indicated in Figure 12E, and the eight indicated in Figure 12F also each sum to 260. The eight squares indicated by the full line of Figure 12G, and those indicated by the dashed line in Figure 12G, also each sum to 260, and a similar remark can be made of the squares indicated by the full and dashed lines of Figure 12H and Figure 12I. We thus have a total of 46 sums of 260 each. Finally, each quarter of the given square is a 4 × 4 semimagic square of magic sum 130.

349° *Dr. Frierson's pandiagonal Franklin magic square.* A magic or a semimagic square each of whose bent diagonals has the magic sum of the square is known as a *Franklin square*. The Franklin square of Item 348° is semimagic, and for some years no one produced a *magic* Franklin square. Over a century after Franklin's introduction of his squares, a Dr. Frierson constructed the 8 × 8 Franklin square shown in Figure 13, and this square is *pandiagonally magic* (that is, not only each row and each column, but also each principal diagonal and each broken diagonal, sums to 260). Dr. Fierson's square also possesses the properties of Figures 12B, 12C, 12D, 12E, 12F, and 12G. Moreover, each quarter of the square is a 4 × 4 pandiagonally magic square.

64	57	4	5	56	49	12	13
3	6	63	58	11	14	55	50
61	60	1	8	53	52	9	16
2	7	62	59	10	15	54	51
48	41	20	21	40	33	28	29
19	22	47	42	27	30	39	34
45	44	17	24	37	36	25	32
18	23	46	43	26	31	38	35

FIGURE   13

350° *A nesting magic square.* Figure 14 shows a magic square of order 7 containing concentrically within it magic squares of orders 5 and 3.

4	9	8	47	48	49	10
38	19	20	17	34	35	12
39	37	26	27	22	13	11
43	36	21	25	29	14	7
6	18	28	23	24	32	44
5	15	30	33	16	31	45
40	41	42	3	2	1	46

FIGURE 14

351° *The IXOHOXI.* The square shown in Figure 15 is a magic square with magic sum 19998 whether the square is viewed rightside up, upside down, or in a mirror. This square, for apparent reasons, is called the IXOHOXI (pronounced iks-ō-hox-ē).

8818	1111	8188	1881
8181	1888	8811	1118
1811	8118	1181	8888
1188	8881	1818	8111

FIGURE 15

A CONCLUDING MISCELLANY

352° *Houseboat.* In Item 270° of *Mathematical Circles Revisited* appears the anecdote about Sidney Cabin and the integration of *d(Cabin)/Cabin,* in which the answer comes out as "log *Cabin"* if the constant of integration is neglected. It is interesting that if the constant of integration is *not* neglected, the answer becomes *"houseboat"*—log *(Cabin)* + *C.*—MICHAEL R. VITALE

353° *Sheldon and infinity.*   C. Sheldon, of the University of Alberta, was very fond of talking about infinity. He pretended to be always on the lookout for it. Toward the end of a lecture he would draw a horizontal line on the blackboard, extend it to the end of the board and beyond, on the wall, toward the door; then he would open the door and continue the line around it and go off to his office. Next day he would return to class with cupped hands and exclaim, "I have it! I have it right here! I finally caught infinity!" Then he would stand near the open window and slowly open his hands. "Oh! Oh!" he would exclaim, "There it goes out of the window. It's gone! It's gone!"

On one occasion at this point, a student who had failed the course the previous year and to whom Sheldon's performance came as no surprise, jumped up, drew a toy pistol from his pocket, aimed the gun out of the window, and fired.

When a new library was being built at the University of Alberta, the excavators had made quite a large mound of earth. Some students mounted a sign on top of the mound which read: MT. SHELDON. YOU CAN SEE INFINITY FROM HERE.

—LEO MOSER

354° *The true founders of the Institute at Princeton.*   Carl Kaysen, a Harvard political economist who succeeded J. Robert Oppenheimer as director of the Institute for Advanced Studies at Princeton, once remarked: "It is not noted in the official history, but the true founders of the Institute for Advanced Studies were Abraham Flexner and Adolph Hitler."

355° *A suggestion.*   Mathematics instructor to an uneasy student: "If you'd rather your parents didn't know your grade, leave a space between your first and last name, and I'll fill in your letter grade as your middle initial."

356° *A lonely profession.* A botanist can easily talk interestingly about his work to a person of another field. The same can be said of a geologist, a physicist, a biologist, an historian, an artist, a musician, an economist, a philosopher, a writer, an astronomer, and so on. But it is pretty difficult for a mathematician to talk mathematics to anyone else but another mathematician. This makes mathematics a lonely profession, and accounts for why mathematicians will travel great distances to attend meetings at which other mathematicians will be present. And the deeper one is in his mathematics, the more difficult it becomes to find someone to talk to. A mathematician at a university, for example, can even be lonely among his colleagues at the university if he happens to be, say, the only mathematical logician there. They tell of a topologist who traveled 2500 miles to meet another topologist so that he might tell of some of the things he had been doing in topology. When he met the other topologist he started right in, telling about his work. The second topologist seemed not to understand anything the first one was saying, and finally blurted out: "Oh, I see, you are a *combinatorial* topologist; I'm a *differential* topologist, you see." And neither one could talk shop to the other. How often a mathematician, discovering an especially attractive proof of a theorem, longs for, but cannot find, someone to whom he can describe his solution.

357° *Overheard at a university.* (a) "My wife doesn't understand me. I'm an algebraic topologist."

(b) Commented a frustrated math professor: "He's the first student the headshrinkers ever failed with."

(c) "I guess I've lost another pupil," sighed the math professor as his glass eye slid down the drain.

358° *Headline.* The headline of a newspaper item read:
UNEMPLOYED MATHEMATICS TEACHER LACKS CLASS.

359° *Point of view.* To illustrate how opinions may vary, depending upon one's point of view, consider the following. A young Roman student Cassius handed to his teacher Brutus an arithmetic problem that looked like this:

which, from the teacher's point of view was wrong. But turn the page upside down and you'll see that from the student's point of view the problem was correct.

360° *A matter of punctuation.* In a now forgotten novel, written in the early part of the present century, occurs the following schoolroom incident:

> "How many one-quarter-inch squares are there in a rectangle of width one inch and length one and a quarter inches?" asked Mr. Fuzzleton of his class. Immediately up shot Scott's arm, and, unable to repress himself, the lad excitedly shouted, "Twenty!"

It can be shown that 20! one-quarter-inch squares are more than enough to cover the entire United States.

L'ENVOI

*From Ebbinghaus' illusion
to equivocal figures*

GEOMETRICAL ILLUSIONS

A GOOD geometrical illusion is clever and amusing, and may be regarded as a pictorial mathematical joke or story. Indeed, a good geometrical illusion, being a condensed pictorial account of an interesting geometrical incident, can be regarded as a special kind of mathematical anecdote. It may, then, be quite proper to include some geometrical illusions in our collection of mathematical stories and anecdotes.

On the principle that it is poor showmanship to explain a joke or a story, we make little effort here to explain the geometrical illusions—we merely present them. For an analysis of why our eyes on occasion deceive us, the reader may consult M. Luckiesh's excellent work, *Optical Illusions,* reprinted by Dover Publications in 1965.

My own fascination in, and wide collection of, geometrical illusions started years ago with Luckiesh's interesting book. Just as it is my hope some day to assemble into a book my collection of mathematically motivated art (a peek at which was offered in *Mathematical Circles Squared*), it is also my hope some day to assemble into a book my collection of geometrical illusions. Here are just a few of them—all simple ones and all involving *circles*.

In looking at geometrical illusions, it must be kept in mind that what may greatly deceive one person's eyes may not so greatly, even if at all, deceive another person's eyes.

In our four trips around the mathematical circle (*In Mathematical Circles, Mathematical Circles Revisited, Mathematical Circles Squared,* and *Mathematical Circles Adieu*) we have related a total of 4(360) = 1440 mathematical stories and anecdotes, plus 9 more in an Addenda to *Mathematical Circles Squared.* So we now begin this final little collection with 1450°.

AREAL DECEPTIONS

1450° *Ebbinghaus' illusion.*   This illusion is a striking example of an areal deception. In Figure 16, though the central circles
in the two parts of the figure are equal, the one surrounded by
small circles appears greater than the one surrounded by large
circles.

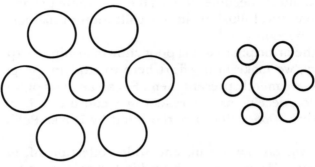

FIGURE   16

1451° *A general principle.*   A generalization of the Ebbinghaus illusion states that, of two equal figures, one adjacent to small
extents and one adjacent to large extents, the former figure appears to be larger. Thus, in Figure 17, though the two interior
circles are equal, the one on the left appears to be larger than the
one on the right. Again, in Figure 18, though the two circles are
equal, the one on the left appears larger than the one on the right.

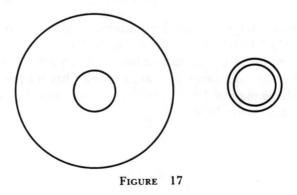

FIGURE   17

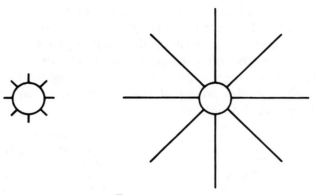

FIGURE   18

1452° *The rising moon.*   Probably everyone has observed, at one time or another, that the sun or full moon when near the horizon appears much larger than when at a higher altitude. That this is only an illusion becomes quite evident if one should view the sun or full moon, when near the horizon, through a cylindrical tube of small diameter; the sun or moon seems instantly to shrink to its size when viewed overhead. Many explanations have been offered to account for the above illusion, one being based upon Figure 19, wherein the circle close to the vertex of the angle appears larger than the other circle, though the two circles are actually equal.

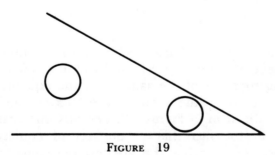

FIGURE   19

1453° *Interior and exterior extents.*   Generally, of two equal convex figures, say, one containing an adjacent extent in its inte-

159

rior and one containing an adjacent extent in its exterior, the latter convex figure will appear to be larger. This is illustrated in Figure 20, in which the interior circle on the right appears larger than the exterior circle on the left, though the two circles are actually equal. The illusion is seen even more strikingly in Figure 21, in which, though the two circles are equal, the one on the left appears larger than the one on the right.

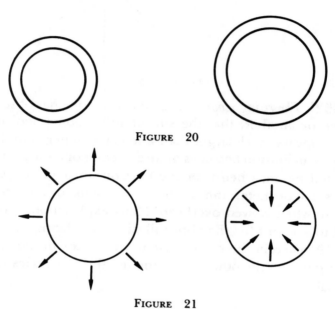

Figure 20

Figure 21

1454° *A property of the circle.* If one should draw, on the same sheet of paper, a circle and the first few regular polygons, making them all of the same *area*, the equilateral triangle would *appear* to have the greatest areal extent and the circle the least. This is seen in Figure 22. Perhaps one reason for this illusion lies in the fact that in our set of equiareal figures, the equilateral triangle has the greatest perimeter and the circle the least.

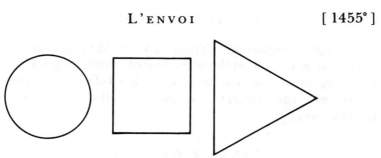

FIGURE   22

1455° *False judgments.*   A judgment of comparison between two areas can sometimes be influenced merely by how the two areas are situated with respect to one another. Thus, in Figure 23, though the two figures are congruent, the lower one appears to be somewhat larger than the upper one. On the other hand, of the two congruent figures in Figure 24, it is now the upper one that appears somewhat larger than the lower one.

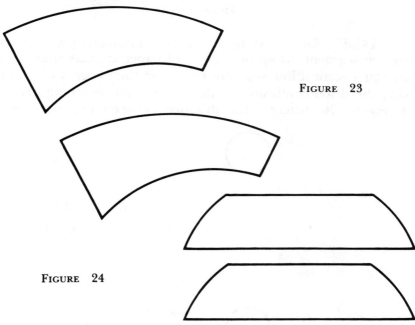

FIGURE   23

FIGURE   24

1456° *Perspective.*   There are striking illusions involving perspective, or at least the influence of converging lines. An example appears in the Frontispiece of this volume, wherein the two "balls" are really equal to one another, but do not appear so to the normal eye.

LINEAR DECEPTIONS

1457° *Three circles.*   In Figure 25, the distance between the inside edges of the two circles on the left is actually equal to the distance between the outside edges of the two circles on the right.

FIGURE   25

1458° *Three circles again.*   To most human eyes, a horizontal line segment drawn on a piece of paper appears shorter than an equal vertical line segment drawn on the paper. Utilizing this fact, the distance illusion of Item 1457° can be strengthened as in Figure 26, wherein the distance between the inside edges

FIGURE   26

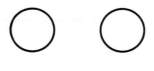

162

of the vertically aligned circles is actually equal to the distance between the outside edges of the two horizontally aligned circles.

1459° *Baldwin's illusion.*   In Figure 27, the little fulcrum is located at the center of the line segment joining the two circles. Though this illusion is not as positive as many others, most viewers would feel that the fulcrum is placed closer to the larger circle.

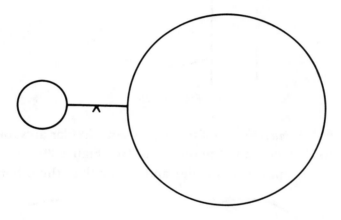

Figure 27

DEFORMATIONS

1460° *An interrupted circumference.*   Most people would feel that in Figure 28, the outer arc on the left of the vertical interruption is the continuation of the rest of the circle, whereas it is actually the inner arc that forms the continuation.

163

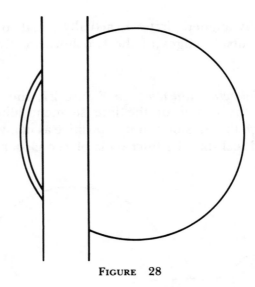

FIGURE   28

1461° *Semicircles.*   Of two equal semicircular arcs, one possessing its diameter and the other not (see Figure 29), the former will generally appear as smaller and flatter than the other.

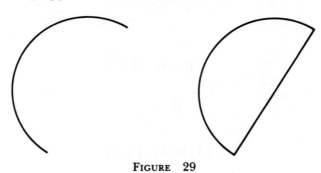

FIGURE   29

1462° *Another interrupted circumference.*   In Figure 30, the short arc of the interrupted circumference of a circle appears flatter and of greater radius of curvature than the larger arcs.

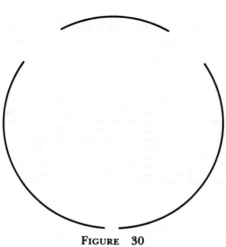

FIGURE 30

1463° *Pinched-in circumference.* To most eyes, a polygon inscribed in a circle appears to distort, or pinch in, the circle at the vertices of the polygon. This is illustrated in Figure 31, where the circumscribed circle appears to dent inward at the vertices of the hexagon. Similarly, the sides of the hexagon appear to sag inward at the points of contact with the inscribed circle.

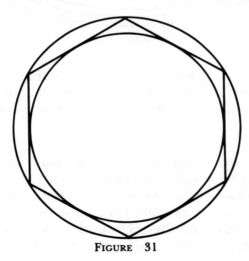

FIGURE 31

1464° *The sagging line.*   The circular arcs above the horizontal straight line segment of Figure 32 cause the segment to appear to sag slightly in the middle.

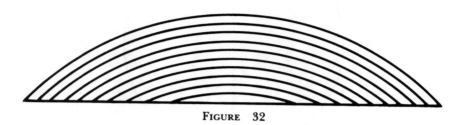

FIGURE   32

1465° *The arching line.*   By placing the circular arcs of Figure 32 below the line segment, as in Figure 33, the horizontal straight line segment now appears to arch slightly in the middle.

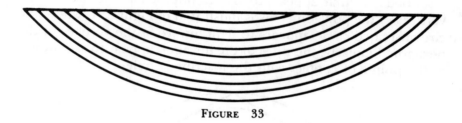

FIGURE   33

1466° *Is it really round?*   Figure 34 strikingly illustrates how a figure can appear quite deformed by merely placing it in an appropriate field. The closed curve of the figure is actually a true circle.

1467° *Equivocal figures.*   There are many geometrical illusions involving what may be called *equivocal figures*; that is, figures that can change in appearance because of the viewer's fluctuation in attention or in association. Thus most of us at one time or another have viewed the famous Schröder reversing staircase, the Thiéry reversible chimneys, Mach's "open book," or the reversible cubes. Thus, an intaglio can sometimes appear as a bas-relief, and vice versa. Figure 35 shows a simple example of an equivocal figure involving a group of rings or circles. Is the "slinky" receding toward the left or receding toward the right? A number of M. C. Escher's tantalizing graphics depend, for their effect, on embedded equivocal figures.

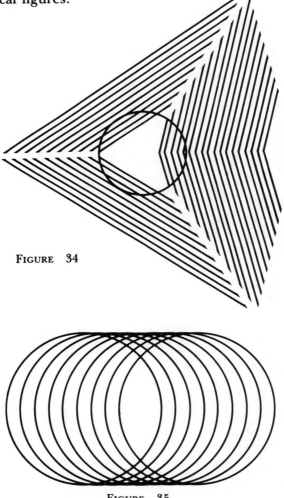

FIGURE   34

FIGURE   35

Return to
Mathematical Circles

Green's Mill at Nottingham, alluded to in item 14°.

Return to
Mathematical Circles

Howard Eves

SELECTED ILLUSTRATIONS
BY CINDY EVES-THOMAS

Published and Distributed by
The Mathematical Association of America

PREFACE

There can be little doubt that Arthur Conan Doyle enjoyed writing his Sherlock Holmes tales, but when he found them interfering with his "more serious" work, he felt it wise to get rid of the beloved detective. Accordingly, in "The Final Problem," at the end of the *Memoirs of Sherlock Holmes,* Doyle managed to have the detective and his arch enemy Professor Moriarty, while locked in mortal combat, presumably topple to their deaths from a ledge high above the great Reichenbach Falls of Switzerland. But Doyle's readers gave the author little peace for this dastardly act, and a number of years later he was forced to revive the famous detective,which he accomplished in *The Return of Sherlock Holmes.*

It is in much the same vein that I now return to mathematical circles, and once again leisurely ramble around the circuit, hoping that by so doing I will appease those who, in so many letters to me, cried foul of my act of some ten years ago, when I tried to get rid of mathematical circles by bidding them an absolute farewell. I recall that, at the time, even the publisher begged me to change the title of the last book from *Mathematical Circles Adieu* to *Mathematical Circles au Revoir.*

So, my good friends, *let* us once more together ramble around the circle. Many thanks to all who sent me stories to be included in a possible new trip. I hope I have nowhere inadvertently failed properly to credit any of the new stories. May *Return to Mathematical Circles* be as useful to teachers at all levels of mathematics as have been the previous circle books.

Howard W. Eves

ACKNOWLEDGMENTS

Appreciative thanks are extended to the following fine journals for graciously allowing reproduction of bits of their material: *American Mathematical Monthly, Journal of Recreational Mathematics, Mathematics Magazine, Mathematics Teacher, School Science and Mathematics, Two-Year College Mathematics Journal* (now *College Mathematics Journal*). And, of course, very special thanks to the writers who are quoted.

CONTENTS

CONTENTS

CONTENTS

QUADRANT TWO

CONTENTS

CONTENTS

QUADRANT THREE

QUADRANT FOUR

RECREATION CORNER

HAVE YOU HEARD?

CONTENTS

EPILOGUE

QUADRANT ONE

*From Newton's original bed
to a universal language*

CONCERNING SOME MEN OF MATHEMATICS

1° *Newton's original bed.*   Isaac Newton's original bed, used by him when he was a youngster, would certainly constitute a fine piece in any mathematical museum. This bed is still in existence and survives today in a very curious place.

In 1948, Roger Babson, best known for his stock market tips, founded the Gravity Research Foundation. The foundation, though interested in any and all kinds of work on gravity, is principally concerned with stimulating searches for some kind of "gravity screen"—that is, for some kind of substance that will cut off the pull of gravity much as a sheet of steel cuts off a beam of light.

Most of this foundation's work has been ludicrous. For example, the foundation has spent several years collecting data on how mental patients are affected by the phases of the moon, since the gravitational pull of the sun and moon may disturb something in either the brain or the spinal fluid. The foundation mailed hundreds of letters to chiefs of police. The purpose of the letters was to find out whether more police calls occur during a full moon. Insurance companies were also contacted to see if accident rates could be correlated with the moon's phases. The foundation did work on gravity and posture. It maintained that if one is to climb a high hill, it is best to do so during a high tide when one's weight is diminished. Perhaps bowing in prayer is recommended because of the change of the direction of the pull of gravity on the brain. Convinced that gravity has an important part to play in ventilation, the foundation suggested that gravity will clear bad air from a building if one gives a slight slant to all the floors, with air outlets at the lower sides; the bad air will slide out through the vents like rain water slides off a sloping roof. The foundation actually built a house in New Boston in which all floors slope a half inch to the foot. The foundation has discussed the effects of gravity on crops, on business, and on political elections. Gravity chairs have been designed to assist in proper circulation of the blood. Priscolene, a patent medicine, is sold by the foundation as an antigravity pill to help circulation.

It was in the fall of 1951 that the foundation held, in New Boston, its first summer conference. On display at the conference was Isaac Newton's original bed, acquired with the monies of the foundation because of the foundation's deep admiration of the discoverer of the universal law of gravitation. Babson's wife became the possessor of one of the world's largest collections of books by and about Isaac Newton.

2° *Plato's shade tree.*   It is claimed that the banyan tree, against which Buddha reclined while meditating, still stands and thrives in India. Though there is some doubt of the authenticity of this, it is more certain that the shade tree under which Plato is said to have lectured to his pupils is still in existence.

Though Plato was not himself a mathematician, he appreciated mathematics and trained many mathematicians in his celebrated academy. He lived between 427 and 347 B.C. and was said to have sought the shade of a fifteen-foot olive tree when he gathered his pupils about him on hot summer days. The tree is one of several in an olive grove but was identified as Plato's tree in 1931 when remains of Plato's academy at Athens were found in nearby excavations. The tree was placed under the protection of the Greek Archeological Service, and in recent times it stood dust-covered by the side of a busy highway connecting Athens with its port city of Piraeus. In 1976 tree experts, in cooperation with the Greek Atomic Energy Center, employed the carbon-dating method to determine that the tree is about three thousand years old.

In October of 1976, a heavy bus crashed off the highway into Plato's tree, breaking the great twisted trunk into four main pieces. The Greek government immediately assigned top priority to saving the ancient tree. Miss Spathari, of the Greek Archeological Service, announced one year after the tragic accident that new shoots had grown from the original massive trunk. The tree now exists only in the form of a bush, protected by a heavy metal barrier to avoid any repetition of the earlier accident. Many years will pass before it reaches anything like its former size and shape.

There is another ancient tree, a plane tree, on the Aegean

4

island of Cos, of about the same age as Plato's tree. This tree also is under the protection of the Greek Archeological Service for it is believed that it was under this tree that Hippocrates of Cos, the classical father of medicine, sat and lectured on the medical practices of his time.

ILLUSTRATION FOR 2°

3° *Hommage à Archimède.*   In the central quad at San Jose State University, directly across from the landmark tower, there now stands a noteworthy seven-foot bronze abstract sculpture, *Hommage à Archimède*, which provides an already pleasant place with an additional pleasant intellectual sweep. Made possible by contributions from friends of the School of Science at the University, the sculpture incorporates several artistic design ideas, a principal one being somewhat reminiscent of the gravestone of Archimedes. Built into the sculpture is a large rectangular figure of "divine proportions," reflecting the observation that if we rotate this figure about its central axis, we find that the resulting volumes satisfy

Cone : Ellipsoid : Cylinder = 1 : 2 : 3.

—LESTER H. LANGE
American Mathematical Monthly, May 1981.

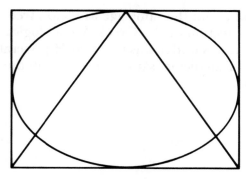

ILLUSTRATION FOR 3°

4° A prophecy. As young men, Eudoxus and Plato traveled together in Egypt and made a stop at Heliopolis, taking up residence with the priests there. During the stay a portentous event came to pass. At some temple service, the sacred bull was observed to lick Eudoxus' garment. According to the priests, this foreboded both good and ill. Eudoxus, they prophesied, would become illustrious but would be short-lived. The discernment of the priests was remarkable; Eudoxus became very illustrious indeed, and he died when he was fifty-three.

5° An impious remark. Anaxagoras was a Greek mathematician and astronomer who lived in the fifth century before Christ, and it is reported that his astronomy was almost the death of him—for he incautiously broached the opinion that the sun is probably but slightly larger in size than the lower end of Greece. He was clapped into prison, condemned to death for impiety, and was saved only by the solicitations of friends.

6° At long last. In 1980, 347 years after Galileo was condemned by the Catholic Church for using his telescopes to prove the earth revolves around the sun, the Vatican, under a call issued by Pope John Paul II, began to review Galileo's conviction of heresy. The review of Galileo's case was called as part of the Church's effort to show that modern science does not negate Christian teaching.

7° *Aryabhata*.   India's first scientific satellite was successfully launched from a Soviet cosmodrome with the help of a Soviet rocket carrier on April 19, 1975 at 1300 hours Indian standard time. The satellite was named Aryabhata, after the famous Indian astronomer and mathematician, who was born in Kusumapara, near present-day Patna, in A.D. 476.

8° *Sir Henry Billingsley*.   The first complete English translation of Euclid's *Elements* was the monumental Billingsley translation issued in 1570. Since so few mathematicians know much about Sir Henry Billingsley, some brief biographical notes may be in order.

> Henry Billingsley was of humble origin, and though he contrived to study for three years at Oxford, he was afterwards apprenticed to a haberdasher. His mathematical learning was acquired mainly from an Augustinian friar, Whytehead by name, who was "put in his shifts" by the dissolution of the monasteries. Being maintained at Billingsley's charges, the friar taught him all his mathematics, and there is no reason to believe that the good patron was a bad student. Billingsley's career was full of great successes; he acquired wealth as well as culture, became Lord Mayor of London, was knighted, and died at a ripe old age in 1606.
>
> —W. B. FRANKLAND
> *The Story of Euclid.* Hodder and Stoughton, 1901.

9° *Philatelists take note*.   To mark the passing of two hundred years since the death of Leonhard Euler and to celebrate Euler's life in Berlin, an East German (DDR) postage stamp and special postmark for Berlin was issued in 1983. The postmark displays two of Euler's famous formulas:

$$e - k + f = 2 \qquad \text{and} \qquad K = \pi^2 EI/l^2.$$

The first of the above formulas would appear in English works as

$$v - e + f = 2,$$

7

where v, e, f represent the number of vertices, edges, and faces of any simple polyhedron. In German, vertices, edges, faces (or surfaces) are called Eckes, Kantes, and Fläches, giving rise to the German formula

$$e - k + f = 2.$$

10° *God versus mathematics.*   A biologist once claimed to have seen a radiolaria covered with a perfect map of hexagons. Upon being informed that Euler had proved this impossible, the biologist replied, "That proves the superiority of God over mathematics."

"Euler's proof happened to be correct," writes Warren S. McCulloch, "and the observation inaccurate. Had both been right, far from proving God's superiority to logic, they would have impugned His wit by catching Him in a contradiction."

11° *Most dogs are nicer than most people.*   Blaise Pascal once commented, "The more I see of men, the better I like my dog."

12° *Crelle's Journal.*   An amusing story is told about the mathematical periodical founded in 1826 by August Leopold Crelle (1780–1855). Though the journal, which is still in existence today, soon became popularly known as *Crelle's Journal*, Crelle named it *Journal für die reine und angewandte Mathematik* (Journal for Pure and Applied Mathematics). Since Crelle was primarily an engineer, it was anticipated that his journal would favor articles on applied mathematics. However, the contrary turned out to be the case, and Crelle published far more articles in pure mathematics than in applied mathematics, accepting in the first year, for example, no less than five long papers submitted by Nils Abel. So some wags suggested that the two words *und angewandte* in the title of the journal should be replaced by the similarly sounding single word *unangewandte*, changing the name of the journal to *Journal für reine unangewandte Mathematik* (Journal for Pure Unapplied Mathematics).

13° *A calculus antagonist.* Michel Rolle (1652–1719) is known to all calculus students for the theorem of beginning calculus that bears his name and says that $f'(x) = 0$ has at least one real root lying between two successive real roots of $f(x) = 0$. Few calculus students, however, know that Rolle was one of the most vocal critics of the calculus and that he strove to demonstrate that the subject gave erroneous results and was based upon unsound reasoning. So vigorous were his quarrels with the calculus that on several occasions the Académie des Sciences felt obliged to intervene.

14° *Miller mathematician.* Another name well known to calculus students is that of George Green (1793–1841), whose famous theorem in calculus books is today basic in theories of electricity and magnetism. But how many calculus students know anything about Green's life?

Green left school, after only one year's attendance, to work in his father's bakery. When the father opened a windmill, the boy used an upper room as a study in which he taught himself physics and mathematics from library books. In 1828, when he was thirty-five years old, he published his most important work, *An Essay on the Application of Mathematical Analysis to the Theory of Electricity and Magnetism.* This article, which received little notice because of poor circulation, contains his famous theorem.

When his father died in 1829, some of George's friends urged him to seek a college education. After four years of self-study during which he closed the gaps in his elementary education, Green was admitted to Caius College of Cambridge University, from which he graduated four years later after a disappointing performance on his final examinations. Later, however, he was appointed Perce Fellow of Caius College. Two years after his appointment he died, and his famous 1828 paper was republished, this time reaching a much wider audience. This paper has been described as "the beginning of mathematical physics in England." In recent years some eight Nobel laureates have used his "functions" in their prize-winning works.

In 1923 the Green windmill was partially restored by a local businessman as a gesture of tribute to Green. Einstein came to pay homage. Then a fire in 1947 destroyed the renovations. Thirty years later the idea of a memorial was once again mooted, and sufficient money was raised to purchase the mill and present it to the sympathetic Nottingham City Council. In 1980 the George Green Memorial Appeal was launched to secure £20,000 to get the sails turning again and the machinery working once more.

ILLUSTRATION FOR 14°

15° *Weaver mathematician.*   Calculus students also meet the name Thomas Simpson in connection with the elegant *Simpson's Rule* for approximating planar areas.

Thomas Simpson (1714–1761) became a noted English mathematician. The son of a weaver, he started working in his father's trade early and consequently received very little formal education as a boy. But in 1724 he witnessed a solar eclipse and secured two books, one on astrology and one on arithmetic, from an itinerant peddler. These two events roused his interest in mathematics. With his newly gained knowledge, he soon became a successful local fortune teller. He further improved his financial situation by marrying his landlady, who was considerably older. In 1733 he settled in Derby where he worked at weaving in the day and teaching school in the evenings. In 1736 he moved to

London and shortly published a highly successful calculus text-book. Now completely emancipated from weaving, he concentrated on teaching and on textbook writing, producing a succession of best-selling textbooks in algebra, geometry, trigonometry, and other areas of mathematics. These books ran into many editions and were translated into a number of foreign languages.

Incidentally, the *Simpson's Rule* that brings Simpson's name to the attention of calculus students was not discovered by Simpson—it was already well known in his time.

16° *Sharing a chair.* Gabriel Cramer (1704–1752) was a Swiss mathematician whose dissemination of ideas of deeper mathematicians earned him a well-deserved place in the history of mathematics. When twenty years old, he competed for the chair of philosophy at the Académie de Calvin in Geneva. Though he failed to secure the chair, the awarding magistrates were so impressed with both Cramer and a fellow competitor that they created a new chair, a chair of mathematics, to be shared by the two men. Each man assumed the full responsibility and salary associated with the chair for two or three years, while the other traveled.

It was during Cramer's travels that he met many of the great mathematicians of his day—the Bernoullis, Euler, D'Alembert, Halley, and others. Eventually, Cramer became the sole occupant of the chair of mathematics and of the chair of philosophy as well.

17° *Newton, the theologian.* A calculus student's usual impression of Isaac Newton is of a great scientist and mathematician, who was guided by unerring logic and worked for the benefit of humanity by unraveling the secrets of the mechanism of the universe.

Actually, Newton claimed that the purpose of his work was to support the existence of God. All his life he worked fervently trying to date biblical events by relating them to various astronomical phenomena. He was so taken with this passion that he frittered away years of his life searching through the Book of Daniel for clues to the end of the world and to the geography of hell.

18° *Bust of Ramanujan unveiled.*   It may surprise more than a few readers that the widow of Srinivasa Ramanujan was still alive in 1986, more than sixty-five years after the famous mathematician's death. Her desire to see her husband memorialized in India by a statue led Richard A. Askey (Wisconsin) to commission a bronze bust and organize a subscription campaign (still open in 1986) to which more than a hundred mathematicians and scientists have contributed. (A sidebar notes possible confusion over Ramanujan's year of birth, most likely occasioned by misreading a numeral in one of his letters; the centenary of his birth will be celebrated in 1987.)

19° *Hardy's lectures on Ramanujan.*   In 1954, G. H. Hardy gave a series of twelve lectures at Harvard devoted to Ramanujan's work. The audience was extremely large. Hardy overcame the difficulty of presenting the necessary formulas by having these formulas numbered and printed in advance and passed out like programs at a concert.

20° *Hardy's most ardent desires.*   G. H. Hardy (1877–1947), one of England's foremost mathematicians and an outstanding expert in analytical number theory, possessed a rebellious spirit. He once listed his four most ardent wishes: (1) to prove the Riemann hypothesis, (2) to make a brilliant play in a crucial cricket match, (3) to prove the nonexistence of God, and (4) to assassinate Benito Mussolini.

In an earlier anecdote on Hardy (Item 108° of *Mathematical Circles Adieu*), we gave Hardy's list of the only important personalities of the world who had achieved a hundred percent of what they wanted to achieve.

21° *Ever a logician.*   Kurt Gödel (1906–1978), the profound Austrian logician and mathematician, mistrusted common sense as a means of discovering truth. It is said that for several years he resisted becoming an American citizen because he found logical contradictions in the Constitution.

22° *An unusual conference.*   It is not uncommon for mathematicians to schedule a conference honoring some great mathematician. Such a conference is often held on a date associated with the year of birth of the mathematician concerned. Thus, a conference was held in 1982 at Bryn Mawr College to celebrate the centennial of the birth of Emmy Noether (1882–1935), who for a time taught at Bryn Mawr, and in 1977 a number of conferences were held honoring the bicentennial of the birth of Karl Friedrich Gauss (1777–1855). In 1979 there were a great many gatherings honoring the centennial of the birth of Albert Einstein (1879–1955).

Quite unusual, however, was a conference held in 1985 at Framingham State College in Framingham, Massachusetts to celebrate the hundredth anniversary of, not the birth of some great mathematician, but the birth of a great mathematical theorem, the *Weierstrass Approximation Theorem:* "Any continuous function over a closed interval on the real axis can be expressed in that intervals as an absolutely and uniformly convergent series of polynomials." This remarkable theorem was published by Karl Weierstrass (1815–1897) in July of 1885, when he was seventy years old, in *Preussische Akademie der Wissenschaften Naturwissenschaftlich Mittheilingen.*

To give due respect to the personal history of Weierstrass, who taught "gymnasium" (high school) by day and did mathematics by night for fifteen years while having no real contact with research mathematicians, some local high school teachers were invited, and the speakers were asked to make their talks suitable for such participants.

The Weierstrass Approximation Theorem aroused enormous interest during the last quarter of the nineteenth century. This interest led to several different proofs and new open problems during the twentieth century. Some of these open problems have been solved; others are still being studied. Thus Weierstrass discovered a theorem that has captured the interest of the mathematical community for one hundred years and is an important part of the body of knowledge known as mathematics. Furthermore, the life of Weierstrass can be a great inspiration to anyone working in mathematics, since he did not obtain his doctorate until he was forty-one years old and since he continued doing mathematics for the rest of his life. In addition, he won the enviable reputation of being not only a great researcher but at the same time an outstanding teacher.

Note: In connection with conferences devoted to some specific mathematical accomplishment rather than to a mathematician, one should mention that, between 1924 and 1949, extensive twenty-place logarithm tables were calculated in England in partial celebration of the tercentenary of the discovery of logarithms. Though John Napier (1550–1617) published his discussion of logarithms in 1614 in a brochure entitled *Mirifici logarithmorum canonis descriptio* (A Description of the Wonderful Law of Logarithms), it was not until 1624 that Henry Briggs (1561–1681) published his *Arithmetica logarithmica,* which contained a fourteen-place table of common logarithms of numbers from 1 to 20,000 and from 90,000 to 100,000. The gap from 20,000 to 90,000 was later filled in, with help, by Adriaen Vlacq (1600–1666), a Dutch bookseller and publisher.

23° *A Christmas card.*   Here is a Harvard story: Osgood, on grounds of dimension, disagreed with Huntington's assertion that mass and W/g are equal. To emphasize his point, he sent Huntington a Christmas card reading, "Merry $X-W/g$."

24° *An evaluation.*   It has been said that the *Principia mathematica* of Russell and Whitehead is the outstanding example of an unreadable masterpiece.

25° *Mother's tongue.*   Otto Neugebauer was born in Innsbruck, Austria in 1899 and received his Ph.D. from Göttingen in 1926. In 1934 he moved to Copenhagen; in 1939 he came to the United States. A mathematician was once surprised when Neugebauer wrote him a letter in English, instead of in his mother's tongue. Neugebauer's explanation was that it was not a question of his mother's tongue but of his secretary's tongue.

26° *An unintended hysteron proteron.* Anyone who lectures a great deal has undoubtedly, at one time or another, become caught in the awkward situation of unintentionally reversing the rational order of some thought. One of these occasions overtook Professor Marston Morse at the close of a long and complex proof that he was presenting. He intended to say, "The conclusion is now obvious," but instead came out with, "The obvious is now concluded."

Errors of this sort, or perhaps the uncomfortable act of becoming inextricably tangled in your syntax, seems almost certain to occur when you have been informed that your lecture is going to be videotaped.

27° *Misprinks.* Klaus Galda, in his review (in *The Two-Year College Mathematics Journal,* Jan. 1980) of Raymond M. Smullyan's *What Is the Name of This Book?,* says: "One of the few drawbacks of *What Is the Name of This Book?* is a relatively large number of misprinks."

28° *Mark Kac.* Mark Kac (Kac is pronounced Katz), a professor of mathematics at the University of Southern California, died of cancer on October 25, 1984, at the age of seventy. He was a pioneer in probability theory and its application to number theory.

Kac was born in Poland and came to the United States in 1938. Before going to USC in 1981, he served for twenty years as a professor of mathematics and theoretical physics at Rockefeller University in New York; before that he was a professor of mathematics at Cornell University. He became one of the foremost mathematicians of his generation and earned election to the American Academy of Arts and Sciences and the National Academy of Sciences.

Kac was a first-rate conversationalist, raconteur, and lecturer, and always had a ready quip. Once he was in the audience when Caltech physicist Richard Feynman was giving a lecture. Feynman, who enjoyed making fun of mathematicians, commented that if mathematics did not exist, physicists could construct it in six days. Kac immediately exclaimed, "That's the time it took God to create the world!" Kac made one of his best-known contributions with Feynman in what is known as the Feynman-Kac formula, which overlaps probability and theoretical physics.

29° *A hefty tome.*   In 1696 a hefty tome was published bearing the imposing title: *New Theory of the Earth.* This book, buttressed by much mathematical detail, purported to be a serious effort to explain the origin of the earth and the solar system.

According to the explanation, it all started in the chaotic motion of the tail of an immense comet. Within this tail, the planets (including the earth) and their satellites slowly assumed shape, all traveling in perfectly circular orbits. In the beginning, the earth did not spin on its axis, and so the early "days" of creation were actually a full year in length. It was not until Adam and Eve ate the forbidden apple that the force of the comet's tail started the earth to rotate, finally culminating in exactly 360 of these rotations, or days, during one revolution of the earth in its orbit. The moon, at this time, revolved about the earth in exactly thirty earth days. The atmosphere of the earth was warm and clear, with moisture so finely distributed that no rainbow could form.

Then, on Friday, November 28, 2349 B.C., Divine Providence sent another comet to visit the earth, this one to serve as an instrument of punishment for the wickedness in the world. Water vapor condensed from the comet's tail and fell upon the earth as torrential rain for forty days and forty nights. However, by diligence and hard work, Noah managed to save his family and an ark full of animals from extinction. Either as a result of the great amount of water acquired by the earth or perhaps because of certain magnetic forces, the earth's orbit was forced into an elliptical shape, increasing the year to its present length of 365 + days. In time the skies cleared, the first rainbow appeared, and the excess water drained into the interior of the earth.

The above theory, closely supporting the Old Testament account of the creation of the world, was bolstered by an abundance of diagrams, erudite footnotes in Greek, and substantiating legends drawn from many cultures.

New Theory of the Earth was written by William Whiston, Newton's successor as professor of mathematics at Cambridge University. The book was well received by the author's colleagues and was highly praised by Newton and by John Locke. Whiston, in addition to being a mathematician, was also a clergyman.

ALBERT EINSTEIN

ALBERT Einstein (1879–1955) is by long odds the most popularly admired scientist of modern times, and consequently many stories and anecdotes about him have been circulated. Also a number of fine biographies of Einstein have been written. Outstanding among· them are the following two recent ones: *'Subtle is the Lord . . . ,' The Science and the Life of Albert Einstein* by Abraham Pais, Oxford University Press, New York, 1982 and *Einstein in America, The Scientist's Conscience in the Age of Hitler and Hiroshima* by Jamie Sayen, Crown Publishers, New York, 1985.

Any Einstein admirer will greatly enjoy reading these two books; they contain many of the ensuing Einstein stories. In the previous trips around the mathematical circle, we have already given a large number of other stories and anecdotes about Einstein.

30° *Lost.*   Einstein wasn't at Princeton very long before he gained the reputation of being an absentminded professor. This reputation grew mainly from stories that he had trouble remembering where he lived. It is said that on two different occasions he asked some passerby for directions to Nassau Hall and explained that the reason he wanted to get to Nassau Hall was that he knew his way home from there. On each occasion, upon being asked his address, he was given a more direct route. Each time he thanked his informant but said he would go to Nassau Hall.

31° *The painted door.*   Perhaps the most familiar anecdote about Einstein, after the one told in one of its variations in Item 65°, concerns the color of Einstein's front door. It seems that

Einstein was so frequently mistaking other people's homes on his street as his own, it was decided to paint his front door a bright red to aid him in finding the right house. This story, which has been doubted by some, has been vouched for by George Olsen of nearby Belle Mead, New Jersey, a handyman who often did odd jobs for the Einsteins.

ILLUSTRATION FOR 31°

32° *Einstein's telephone.*   In an effort to attain a degree of privacy, Einstein had an unlisted phone number for his home. Unfortunately, he could never remember the number, so when he would forget where he lived he was unable to call home to find out. It's a matter of record that Einstein once called the Princeton University switchboard to find out where he lived.

33° *A discarded photograph.*   A story has made the rounds about an embarrassing incident that occurred to Einstein one day when he was walking home, lost in deep thought. It seems that his path chanced upon the site of an open street excavation, and before he realized the situation he fell into the hole. The local Princeton photographer, Alan Richards, who happened to be on the spot, snapped a picture of the disaster.

When, with the help of Richards, Einstein climbed out of the hole, abashed but uninjured, he begged Richards not to print the picture. Richards graciously handed Einstein the film.

Einstein and Richards became strong friends over the ensuing years, and the great scientist was often the subject of more complimentary pictures taken by Richards.

The above story reminds one of a similar embarrassing incident that occurred to Thales one night when he was walking along observing the stars. See Item 27° of *In Mathematical Circles.*

34° *Sailing.*   Einstein found great pleasure in sailing. It was not racing or long trips that appealed to him, but rather the enjoyment of being a part of nature and of dreaming and thinking as the wind carried him along. He also enjoyed sailing as an exercise in practical physics. In America he found much opportunity to sail his little boat *Tinef* (roughly meaning "worthless") in both fresh and salt water. But his close friends often worried when he went sailing, for Einstein could not swim. Once, in the summer of 1944, he had a close call. He was sailing with some companions in choppy water when his boat struck a rock, quickly filled with water, and capsized. Einstein got caught in the sail and was under water for quite some time before he managed to free his leg from a rope. Luckily the water was warm, and a motorboat soon rescued the sailors. Throughout the adventure, Einstein's pipe never left his hand.

This story reminds one of Gauss's narrow escape from drowning when he was a small child. See Item 320° of *In Mathematical Circles*.

35° *Einstein as a dyslexic.*   There is a complex childhood disorder known as *dyslexia,* in which the mind involuntarily transforms letters and scrambles words. In reading, dyslexics frequently lose their place and skip a word or a whole line. Realizing the words are not making sense, they will reread the material again and again, in great frustration. The print may appear blurred or in motion, and the letters often appear in reverse order. In mathematics, dyslexics have trouble with rote memory, such as the addition and multiplication tables, but they can understand mathematical concepts.

In Item 116° of *Mathematical Circles Adieu,* we reported that Einstein was a late bloomer, unable or unwilling to speak until he was three years old. In elementary school he had great difficulty with sums and had to be taught the multiplication tables by raps on his knuckles. Dr. Harold N. Levinson, associate professor of psychiatry at New York University Medical Center, has done research on dyslexia. He believes that Einstein's difficulty with elementary mathematics stemmed from the fact that as a youngster he was a dyslexic, although later he triumphantly dealt with vast mathematical concepts.

Among people that Levinson has found to be dyslexics are many highly talented individuals, such as artists, writers, poets, scientists, physicians. "Indeed," says Levinson, "in some individuals, were it not for the underlying dyslexia, their struggles would not have led them to success and fame. It was actually a stimulus to success."

36° *Absentminded professor.*   For many years Einstein complained of pains in his stomach. Doctors finally diagnosed the trouble as arising from malnutrition caused by Einstein's forgetting to eat.

His doctor friend Janos Plesch once noted, "As his mind knows no limits, so his body follows no set rules. He sleeps until awakened; he stays awake until he is told to go to bed; he will go hungry until he is given something to eat; and then he eats until he is stopped."

37° *Trying to catch vibes.* A Princeton University freshman student, who was doing poorly in science, awaited one morning after a fresh fall of snow for Einstein to pass on his way to Fine Hall. Then, a dozen paces behind the great scientist, the student plodded along carefully placing his feet in Einstein's tracks. He had a test in science coming up that day and hoped to improve his chances.

38° *A tense moment.* Hans Panofski once chauffeured his father and Einstein to an art show. Returning the two men home after the show, the boy, who was driving with an expired California driver's license, missed a couple of heartbeats when a policeman stopped the car. To the boy's relief, the policeman had stopped the car simply to assure himself that it was indeed the great scientist that he had spotted passing by.

39° *A bust of Einstein.* "When I first saw Albert Einstein, his body seemed suspended from his head. His hair looked like a spiral nebula." The year was 1953, the place, Einstein's home in Princeton, N. J., where sculptor Robert Berks was working on a bust of the father of atomic science. "The world needs heroes and it's better they be harmless men like me than villains like Hitler," said Einstein, who pondered theories of electromagnetism while Berks sculpted. The artist spent twenty-four years in search of a sponsor to turn a figurine into a full-scale memorial. But now his Einstein, in sweat shirt and sandals, has found a home in Washington. The memorial was commissioned for $1.6 million by the National Academy of Sciences to commemorate Einstein's centennial. Berks says he hopes his twelve-foot statue, cast in bronze, will show the humanitarian side of Einstein. "His strength and gentleness made me see that heroes don't just live in novels."

40° *Absence of adulation.*   George Olsen has told a pretty little tale of an occasion when Einstein did not receive the usual adulation that generally followed him.

Olsen often worked evenings as an usher at the Princeton Playhouse movie theater, which was frequented by Einstein and his daughter. One evening Olsen noted a large expectant crowd gathering just outside the theater and thought that the people had come to give Einstein a reception when he and his daughter emerged from the theater. To his astonishment, when Einstein came out no one seemed even to notice him.

A little later the mystery was explained, when pop singer Johnnie Ray, who was visiting Princeton that evening, emerged amid a frenzy of screams. Einstein disappeared down the street, unaware of the scene taking place behind him.

41° *Disdaining adulation.*   Albert Cantril, son of the psychologist Hadley Cantril, tells of a time Einstein disdained adulation with considerable vehemence. Albert's mother had called at the Einstein residence to obtain autographs in some books for her young children.

As she was explaining the purpose of her visit to Einstein's secretary and was extolling Einstein as a very great man, the scientist himself came downstairs. Overhearing the conversation, Einstein, quite out of usual character, exploded in anger, shouting that adulation is the cult of personality and the breeder of Hitlers and Stalins.

A visiting colleague of Einstein's, who happened to be present at the time, interceded in an effort to smooth over the embarrassing scene and asked Mrs. Cantril to leave the books. The discomfited Mrs. Cantril left, "feeling about two inches high." Later Einstein autographed the books and returned them to Mrs. Cantril.

42° *A contrast.*   Thomas Mann, the eminent German novelist and refugee, resided at Princeton from 1938 to 1941. He found Princeton repugnant and described his lectures there as two years of jokes. He lived like a patrician in a great, red brick bastion on Library Place. Einstein, an eminent German scientist and refugee, also resided in Princeton at the time. He found Princeton a pleasant town, and he enjoyed the occasional lectures that he delivered. He lived a simple life in a small, modest, wood frame house, a block away from Mann's impressive villa.

At Christmas time a group of carolers visited both homes. When they sang before Einstein's home, the scientist came out on the porch in his shirt sleeves and with no socks. As he attentively listened to the carolers, his secretary came out and gently placed a coat over his shoulders. At the conclusion of the singing, he thanked the carolers and shook each of their hands.

Later the same evening, the carolers sang before Mann's imposing home. Costly drapes were drawn across the windows. No one offered thanks. Occasionally, through the cracks in the drapes, ignoring guests could be seen moving about inside as an elegant party was in progress.

On another occasion when a group of Christmas carolers sang to Einstein, he came to the door and smiled. He then walked back into his house, got his violin, and accompanied the group on the rest of their rounds.

43° *A record?*

NEW YORK—An autographed 12-page manuscript by Albert Einstein was sold at an auction Saturday for $55,000, an amount believed to be the most ever paid for any of the late scientist's papers.

The German-language manuscript, which explains Einstein's unified field theory and its place in the history of physics, was sold to M. F. Neville Rare Books of Santa Barbara, Calif., during an auction at Christies.

—Associated Press, *The Orlando Sentinel,*
Sunday, Dec. 18, 1983.

In view of the next Item, one questions the final part of the opening sentence of the above story.

44° *Auctioning a manuscript.* During World War I an organization was formed called The Book and Author War Bond Committee. The committee's purpose was the collection of famous original manuscripts to be auctioned off for war bonds. The committee asked Einstein for the original manuscript of his renowned 1905 paper on special relativity. Einstein replied that the manuscript was no longer in existence, that he had thrown it away after the paper was published. As a substitute, he offered the manuscript of his latest paper, "Bivector Fields" (coauthored by Valentine Bargmann). The committee gratefully accepted this manuscript and, with some hesitancy, suggested that Einstein might copy out in longhand the printed 1905 paper. Einstein acquiesced, and in 1944 the two manuscripts, neatly bound together in a slipcase, fetched $11.5 million at an auction in Kansas City and were then immediately donated by the purchaser to the Library of Congress. Einstein commented, "The economists will have to revise their theories of value."

45° *The Nehru azalea.* India's Prime Minister Jawaharlal Nehru visited Einstein's home on November 5, 1949. The two men had been long time admirers of one another. A few days after the visit, a thank-you gift from Nehru was delivered to Einstein's home at 112 Mercer Street in Princeton. The gift was an azalea, which was duly planted in front of the house and was ever after referred to as the Nehru azalea.

Illustration for 45°

46° *Cross-purposes.* In 1946, Claude Pepper, then a Florida senator, spoke at Princeton University about the conditions in Eastern Europe. After the address, Einstein invited Senator Pepper to his home, for he was eager to learn further about the situation in Eastern Europe. But Pepper was more eager to learn something about relativity theory. For a while the two questioned each other from their disparate points of interest. Finally Pepper won out, and Einstein described his early introduction to relativity.

47° *A favorite short story.* Einstein's favorite short story was Tolstoy's "How Much Earth Does a Man Need?" The story is a charming parable concerning a man whom the devil gives an opportunity to possess all the land the man can walk around in a single day. Because of his greed and imprudence, the man perishes before the day is out, thus furnishing Tolstoy with the answer to the question posed in the story's title—about six feet by two feet, or enough for a grave.

48° *An antidote to tiredness.* Shortly after undergoing a serious operation, Einstein attended a meeting at which many physics lectures were delivered. At the conclusion of the meeting, someone expressed surprise to Einstein that he was not more tired, especially since he was recuperating from surgery. Smiling, Einstein replied, "I would be tired if I had understood them all."

49° *A great grief.* Einstein's younger son, Eduard, suffered from schizophrenia and spent most of his adult life in a home in Switzerland. Einstein and his first wife separated when the extremely sensitive boy was only four years old. Leaving his sons proved to be the most painful experience of Einstein's life, and the subsequent illness of Eduard compounded his grief.

50° *Einstein's brain.* In Item 44° of *Mathematical Circles Adieu,* we reported that the brains of Gauss and Dirichlet are preserved in the department of physiology at Göttingen University. Albert Einstein's brain was removed during an autopsy following the scientist's death in 1955. Einstein died of an aneurysm in the hospital in Princeton, N. J., and his brain was extracted for study in an effort to find clues to his genius. The whereabouts of the brain had been unknown to the public since the autopsy until recently, when portions of the brain were traced to a laboratory in Wichita, Kansas, where "they are floating in a Mason jar."

51° *A code word.*   Some families have code words to warn a member of the family that "You're drinking too much," "You're talking too much," "Your slip is showing," or "It's time to go home."

It seems that Einstein's name has become a code word for the embarrassing situation when someone has neglected to close his trouser zipper. Professor Einstein was a rather careless and casual dresser, usually appearing in unpressed slacks, old sweater or sweat shirt, and often without socks. In his disinterest in the matter of dress, he often failed to close his zipper. Therefore, when a member of a family needs to be warned of this situation, another member of the family will say "Einstein."

Some other families employ, for the same purpose, the mathematical-sounding code *"XYZ,"* standing for "Examine your zipper."

EINSTEIN'S THEORY OF RELATIVITY

52° *A simple explanation.*   One of the duties of Helen Dukas, Einstein's secretary/housekeeper, was to shield the professor from the public. As an intermediary she was often asked about Einstein's scientific work, and the questions were not always easy to answer. Accordingly, if asked to explain relativity, Einstein instructed her to say: "An hour sitting with a pretty girl on a park bench passes like a minute, but a minute sitting on a hot stove seems like an hour."

To this a listener might reply, "From such nonsense Einstein makes a living?"

53° *The big brain.*   A supposed "big brain" was addressing a group, on the subject of Einstein's "relativity." After discussing it for an hour, one of the group halted the speaker: "You know,

Sam, you are greater than Einstein on even his own theory. They claim there are only twelve people in the whole world that understand Einstein and his 'relativity'—but *nobody* understands you!"
—HARRY HERSHFIELD

54° *Down with Einstein.* The eminent British physicist, Oliver Heaviside (1850–1925), was a curious blend of scientist and eccentric. He was famous for his work in electromagnetic theory and for his operational calculus. Increasing deafness, occurring even before he was twenty-five, led him to a withdrawn life. Among his foresights was his advocacy of the more supple vector analysis over the ponderous quaternionic algebra of Hamilton; among his blind spots was his denunciation of Einstein and the theory of relativity. He was the only first-rate physicist at the time to impugn Einstein, and his invectives against relativity theory often bordered on the absurd.

55° *The two buckets.* There are many stories concerning "relativity" that narrators feel must somehow reflect Einstein's theory. Thus there is a fable about two buckets on their way to the well. One commented, "Isn't this uselessness of our being filled depressing? For though we go away full, we always come back empty."

"Dear me! How strange to look at it that way," said the other bucket. "I enjoy the thought that, however empty we come, we always go away full."

56° *Many things are relative.* The difference between a groove and a grave is only a matter of depth.

57° *It's all relative.* A reporter asked a Chinese delegate to the United Nations, "What strikes you as the oddest thing about Americans?" His reply was, "I think it is the peculiar slant of their eyes."

58° *No difference.* Driving in the country one day, a man saw an old fellow sitting on a fence rail, watching the automobiles go by. Stopping to talk, the traveler said, "I never could stand

27

living out here. You don't see anything. You don't travel like I do. I'm going all the time."

The old man on the fence looked down at the stranger slowly and then drawled, "I can't see much difference in what I'm doing and what you're doing. I set on the fence and watch the autos go by and you set in your auto and watch the fences go by. It's just the way you look at things."

59° *Relative density.*   You can send a message around the world in one-seventh of a second, but it may take years to force a simple idea through a quarter-inch human skull.

60° *Everything is relative.*   If a monkey had fallen from the tree in place of an apple, Newton would have discovered the origin of the species instead of the law of gravity.

EINSTEIN AND CHILDREN

61° *A useful accomplishment.*   Einstein was fond of children, and he enjoyed amusing them by wiggling his ears.

62° *An explanation.*   Children often asked Einstein why he didn't wear socks. One little girl said, "Your mother will be afraid you'll catch cold." To some small boys he once explained, "I've reached an age where if somebody tells me to wear socks, I don't have to."

63° *His hair.*   A child seeing Einstein pass by on the street asked his mother, "Is that Mrs. Einstein?" Another youngster, who, on a visit, had been taken upstairs to say hello to Einstein, rushed back downstairs loudly proclaiming, "Mom, you are right, he *does* look like a lion."

64° *Chicken pox.*   Mrs. Wigner dropped off a package for Einstein, who inquired after her children. She replied that they were quite all right except they had the chicken pox. "Where are they?" Einstein asked. "They are waiting outside in the car," she replied. "Oh, I've had that disease," Einstein laughed, and skipped outside to visit with the children.

28

65° *The true account.*   A commonly circulated story about Einstein concerns a little girl who asked the great scientist to help her with her arithmetic, which, the story says, he did, much to the improvement of the little girl's arithmetic marks. When later queried by the little girl's mother as to what he got out of it, Einstein is said to have exclaimed, "Why, every time I help her she gives me a lollipop." (See, for example, Item 97° in *Mathematical Circles Squared.*) Like so many stories about great people, this one has a basis but became somewhat distorted. The true account of the little girl and her arithmetic is as follows.

In the late 1930s, Adelaide Delong was in the third or fourth grade at Miss Fine's School in Princeton. The Delongs lived on the same street as did Einstein, but about a half mile beyond. One day, in passing Einstein's home, Mrs. Delong pointed it out to her young daughter and remarked that the world's greatest mathematician lived there. Now Adelaide was having trouble with her arithmetic, and so one afternoon, instead of returning directly home from school, she called at Einstein's house. Being but an eight-year old child, she was admitted and Einstein was called; he was offered a handful of fudge by the little girl. He accepted the fudge, but upon being asked if he would help her with her arithmetic, he gently declined on the grounds that he felt to do so would be unfair to Adelaide's teacher and the other pupils in the school. In return for the fudge, Einstein gave Adelaide some cookies. Though her mission for help in her arithmetic failed, Adelaide became a favorite with the ladies at Einstein's house, and she became a regular Sunday visitor.

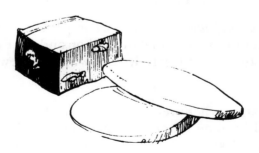

ILLUSTRATION FOR 65°

66° *Quick on the draw.*   Two young boys found that Einstein possessed prowess with a water pistol. Each summer, they would provide themselves and Einstein with squirt guns. It has been reported that Einstein was a good shot both in pace-and-turn dueling and in cowboy-style straight draw. Passersby were often startled and amused to see Einstein and the two boys racing from tree shelter to tree shelter, taking potshots at one another.

67° *Childlike sense of play.*   During Einstein's latter years he was befriended by a number of youngsters. There was one boy who met the great scientist each day on Einstein's return home from his office in Fine Hall at Princeton University. After some banter and an exchange of jokes, the boy was usually taken into the Einstein home to have another look at the scientist's small but impressive chemistry laboratory.

68° *Some correspondence.*   Einstein received a lot of mail from children. A girl in South Africa once wrote that she was surprised he was still alive, since she thought he was a contemporary of Newton. Einstein replied, "I have to apologize to you that I am still among the living. There will be a remedy for this, however." Soon a second letter followed from the girl, confessing that she was a girl and not a boy as Einstein had mistakenly assumed. Einstein responded, "I do not mind that you are a girl. But the main thing is that you yourself do not mind. There is no reason to."

EINSTEIN'S HUMOR

69° *A new definition.*   Once while patiently sitting through a long after-dinner speech that droned on and on, Einstein turned to his neighbor and whispered, "I now have a new definition of infinity."

70° *Tiger.*   Whenever it rained, the Einstein family's cat, Tiger, would become miserable. Einstein would sympathetically say to the cat, "I know what's wrong, my dear, but I really don't know how to turn it off."

71° *Moses.*   Einstein's daughter, Margot, upon returning home from a visit where she had met a large long-haired dog named Moses, concluded her description of the dog to her father by proclaiming that the dog had so much hair that one could not tell his front from his rear. To this Einstein responded, "The main thing is that *he* knows."

72° *Chico.*   Margot agreed to take care of a white-haired terrier named Chico, who soon became a permanent and valued member of the household. Chico developed a special fondness for Einstein. "He is an intelligent dog," commented Einstein, "and sympathizes with me on the matter of my daily inundation with correspondence. He tries to bite the mailman because he brings me too many letters."

73° *Advice.*   Einstein was once asked if he believed it is permissible for a Jew to marry out of his faith. With a hearty laugh he replied, "It's dangerous, but then *all* marriages are dangerous."

EINSTEIN QUOTES AND COMMENTS

74° *More Einstein quotes.*   In the earlier trips around the mathematical circles, we gave a number of pithy Einstein quotes. Here are some more:

> Imagination is more important than knowledge.
> It is nothing short of a miracle that modern methods of instruction have not yet entirely strangled the holy curiosity of inquiry.
> Although I am a typical loner in daily life, my consciousness of belonging to the invisible community of those who strive for truth, beauty, and justice has preserved me from feeling isolated.

75° *An Einstein calendar.*   There appeared an elegant calendar for the year 1985, containing beautiful photographs of Einstein taken by Lotte Jacobi. There was a photograph for each month of the year, and each photograph was accompanied by a pithy Einstein quotation. Here are the quotations:

> The state has become a modern idol whose suggestive power few men are able to escape.

Everything that is really great and inspiring is created by the individual who can labor in freedom.

It is a precarious undertaking to say anything reliable about aims and intentions.

All means prove but a blunt instrument if they have not behind them a living spirit.

One can organize to apply a discovery already made, but not to make one.

We must overcome the horrible obstacles of national frontiers. Security is indivisible.

Perhaps it is an idle task to judge in times when action counts.

Since I do not foresee that atomic energy is to be a great boon for a long time, I have to say that for the present it is a menace.

We know a few things that the politicians do not know.

The road to perdition has ever been accompanied by lip service to an ideal.

I live in that solitude which is painful in youth, but delicious in the years of maturity.

76° *On Goethe.*   Though Einstein admired Goethe's wisdom and cleverness, he did not like the writer's prose. The reason reveals something of the scientist's own character: "I feel in him a certain condescending attitude toward the reader, a certain lack of humble devotion, which, especially in great men, has such a comforting effect."

77° *On convincing one's colleagues.*   The noted psychologist Hadley Cantril, a friend and neighbor of Einstein, once showed Einstein a model of his later-famous trapezoidal room. Einstein instantly saw the psychological and philosophical significance of the illusions created by the apparently square interior.

When Cantril complained that he was disappointed by the disinterest in his findings by fellow psychologists, Einstein consoled his friend by remarking, "I have learned not to waste time trying to convince your colleagues."

78° *Subtle is the Lord.*   Einstein once remarked, "Subtle is the Lord, but malicious He is not." Oswald Veblen, then a member of the mathematics faculty at Princeton University, heard the remark and wrote it down. In 1930, when Fine Hall was built at

Princeton University to house the mathematics department, Veblen secured Einstein's permission to inscribe the statement in the marble above the fireplace in the faculty lounge. Einstein explained that by his statement he meant, "Nature hides her secrets because of her essential loftiness, but not by means of ruse."

79° *Contemplating a death.*   After the death of his sister, Maja (on June 25, 1951), Einstein sat quietly on his back porch with his daughter, Margot. After sitting thus for some time, he pointed to the trees and the sky and gently said to Margot, "Look into nature, then you will understand it better."

80° *On racism.*   When Marian Anderson, the gifted contralto singer, was refused a room at the Nassau Inn in Princeton because of her color, Einstein invited her to stay at his home. Thereafter, on subsequent engagements in Princeton, Marian Anderson always stayed at 112 Mercer Street.

Einstein once commented that the "worst disease" in American society is "the treatment of the Negro. Everyone who is not used from childhood to this injustice suffers from the mere observation. Everyone who freshly learns of this state of affairs at a maturer age, feels not only the injustice, but the scorn of the principle of the Fathers who founded the United States that 'all men are created equal.' " On the same point, he later wrote, "The more I feel an American, the more this situation pains me. I can escape the feeling of complicity in it only by speaking out."

81° *Einstein's formula for success.*   Einstein said his formula for success was: $x + y + z =$ SUCCESS, where x stands for hard work and y stands for play. He was always asked, "Well, what does the z stand for?" "Oh, that," he would reply, surprised, "represents when to listen."

82° *Golden silence.*   One time Einstein and his daughter, Margot, were dining alone. Margot saw that her father was lost in thought as he mechanically ate his dinner, and so she, too, kept quiet. At the end of the silent meal, Einstein looked up at Margot and softly remarked, "Ist dies nicht schön?" (Is this not beautiful?)

33

83° *Really?*   Once, when Einstein's secretary was reading to the scientist from one of his early papers, he interrupted and asked, "Did I really say that?" Upon being assured that he had, he remarked, "I could have said it so much more simply."

84° *Concentration Camp.*   Abraham Flexner, the director of the Institute for Advanced Study at Princeton, so carefully shielded Einstein from visitors and public requests that at one time Einstein became annoyed and so informed his friend Rabbi Wise. The return address on his letter read, "Concentration Camp, Princeton."

LOBACHEVSKI AND JÁNOS BOLYAI

85° *Helping one up the ladder.*   Lobachevski took a deep paternal interest in the young, and stories are told of how he assisted young men in their education. He once observed a young clerk seizing moments to read a mathematics book behind a counter. Lobachevski procured the young man's admission to school, from which the student proceeded to the university and eventually succeeded in occupying the chair of physics there.

Another story tells of a poor priest's son who traveled afoot all the way from Siberia, arriving at Kasan in a destitute condition. Lobachevski took him under his charge and secured the young man's entrance into the medical school at the university. Time showed that the favor was not misplaced, for the student graduated, became a dedicated doctor, and evinced his gratitude by bequeathing a valuable library to the university whose kind rector had so helped him in his time of need.

86° *Lobachevski as a lover of nature.*   At some distance from Kasan, up the Volga, is a little village where the care of gardens and orchards occupied much of Lobachevski's leisure. A melancholy story tells how he planted there a grove of nut trees but had a strong premonition that he would never eat their fruit; and the trees first bore fruit soon after his death. All sorts of agricultural, horticultural, and pastoral matters excited his lively interest, and

his activity in these pursuits was recognized by a silver medal from
the Moscow Imperial Agricultural Society.

—W. B. FRANKLAND
The Story of Euclid. Hodder and Stoughton, 1901.

87° *A comparison.*   If Lobachevski's genius is admirable, that
of János Bolyai is astounding. What the former did by the continued effort of an ample lifetime, the latter seemed to achieve
in one flight of the mind. Lobachevski's tenacity may suggest the
steady growth of a planet; Bolyai's career is that of a brilliant but
transient meteor.

—W. B. FRANKLAND
The Story of Euclid. Hodder and Stoughton, 1901.

88° *A sequel to a well-known story.*   The most frequently
told story about János Bolyai concerns the succession of duels he
fought with thirteen of his brother officers. As a consequence of
some friction, these thirteen officers simultaneously challenged
János, who accepted with the proviso that between duels he should
be permitted to play a short piece on his violin. The concession
granted, he vanquished in turn all thirteen of his opponents. What
is seldom told is what happened very shortly after the batch of
duels. János was promoted to a captaincy on the condition that
he immediately retire with the pension assigned to his new rank.
The government felt bound to consult its interests, for it could
hardly suffer the possibility of such an event recurring.

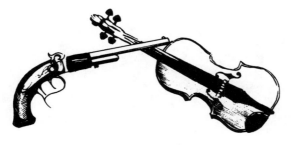

ILLUSTRATION FOR 88°

89° *János Bolyai's disposition.*   The disposition of János Bo-lyai is described as retiring: he lived almost the life of a hermit. Those who encountered him were struck by a strangeness in his ways of acting and thinking—superficial eccentricities for which his geometrical work amply atones, for in the long run the gen-uineness of the gold tells, and its grotesque stamp is forgotten.

—W. B. FRANKLAND
The Story of Euclid. Hodder and Stoughton, 1901.

90° *A universal language.*   Toward the end of his life, János Bolyai concerned himself with the idea of constructing a universal language, claiming what was an accomplished fact in music surely was not beyond hope in other departments of human life.

QUADRANT TWO

From probability
to Dolbear's law

JULIAN LOWELL COOLIDGE

JULIAN Lowell Coolidge (1873–1954), a prominent member of the Harvard mathematics faculty in the first half of the twentieth century, was a descendant of Thomas Jefferson, a cousin of Abbott Lawrence Lowell (the mathematically inclined president of Harvard from 1908 to 1933), a one-time teacher of Franklin Delano Roosevelt (at Groton), a Rough Rider under Theodore Roosevelt in the war of 1898, and a mathematics student of Kowalewski, Study, and Segre from 1902 to 1904. He was elected president of the Mathematical Association of America in 1925 and founded the association's Chauvenet Prize in that year. He was appointed the first master of Lowell House at Harvard in 1930, and over the years he authored a number of remarkable mathematics books (all published by the Oxford University Press).

Coolidge was, in his time, perhaps the foremost geometer on the American continent. The present writer, while a graduate student at Harvard from 1934 to 1936, had the honor and pleasure of working under him in geometrical research. Coolidge was an extraordinary mentor. He possessed a charming wit and sense of humor and had a head crammed with an incredible stock of geometrical knowledge.

91° *Probability.* Coolidge had an interest in the theory of probability. His 1909 paper, "The Gambler's Ruin," was an early investigation of the effect of finite stakes on the prospects of a gambler. He proved, under his assumptions, that the best strategy is to bet the entire stake available on the first turn of a fair coin. "It is true," he concluded, "that a man who does this is a fool. I have only proved that a man who does anything else is an even bigger fool."

92° *Calculating.* Coolidge found pleasure in performing long calculations. This was in the days before the advent of computers, and when the calculations became too extensive for ordinary paper, Coolidge used wallpaper.

93° *Wit and humor.* Coolidge's lectures were often enlivened with wit and humor. In explaining the concept of passing to the limit, Coolidge once stated, "The logarithm function approaches infinity with the argument, but very reluctantly."

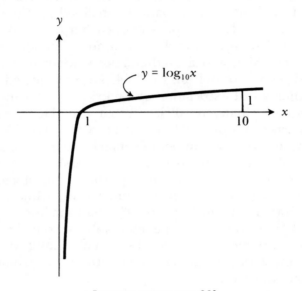

Illustration for 93°

94° *The watch episode.* A frequently recalled anecdote about Coolidge occurred one day during his analytic geometry class. Coolidge had the habit of twirling his gold watch and chain, back and forth around his index finger as he lectured. At the start of a particularly vigorous swing, the chain broke, and the valuable watch looped across the room and landed on a window sill. Coolidge immediately seized the lesson involved and uttered, "That, gentlemen, was a perfect parabola."

95° *Sandy.* There was a relaxed atmosphere in the meetings of the Harvard graduate mathematics club that met once a month. Coolidge frequently brought his large Airedale, Sandy, to these gatherings. Sandy habitually chose a position on the rug

40

directly in front of the middle of the blackboard, and there he would lie quietly during the ensuing talk. The speaker had free use of the far ends of the blackboard, but the center portion was preempted by Sandy.

96° *An observation.*   Coolidge once remarked to one of his graduate research students that a nice thing about mathematics is that it never solves a problem without creating new ones.

SOME MORE STORIES ABOUT MEN OF MATHEMATICS

HERE are a few more stories about eminent mathematicians. The stories in Items 101° to 105° are adapted from tales told by George Pólya to various audiences at various times.

97° *Some simple mathomagic.*   Edward Kasner (1878–1955), one-time Adrian Professor of Mathematics at Columbia University, had a remarkable rapport with children, and he would often visit the grade schools of New York City to talk "mathematics" with the young pupils.

One of Professor Kasner's devices was to employ simple examples of mathomagic. For example, he would display a dime and a nickle and would ask one of the pupils to take the coins and put his hands behind his back with the dime in one hand and the nickle in the other, but not to tell which. Professor Kasner would then say, "I'm going to tell you in which hand you have the nickle and in which hand you have the dime and at the same time test you in some simple arithmetic. Choose a partner to help you check the arithmetic." A partner would be selected.

"Now," Professor Kasner would continue, "without telling me, multiply the value of the coin in your right hand by four and the value of the coin in your left hand by seven, add the two answers, and tell me the result." The calculation would be secretly performed by the two pupils and the result announced. If the result was an even number Professor Kasner would say, "The nickle is in your right hand and the dime is in your left hand."

If the announced result was odd, Professor Kasner would say, "The nickle is in your left hand and the dime is in your right hand."

The pupils would be astounded, and the trick would be repeated with other pairs of pupils until perhaps someone "caught on." By the end of the session, the pupils had had practice in simple multiplication and addition, and at the same time they had learned some basic properties of odd and even numbers.

ILLUSTRATION FOR 97°

98° *A treasure hunt.*   Another of Professor Kasner's devices when talking "mathematics" to grade school children was to have a treasure hunt. He would first entertainingly instruct the class in the basics of some topic, say the concept of prime and composite numbers. Then, taking a position at a far end of the blackboard, he would announce that they were going to have a treasure hunt and that finding the treasure would depend on the class correctly answering a few questions.

The questions would proceed, such as, "What is the second odd prime?" "What is the largest prime factor of 12?", and so on. As the class gave the answers, Professor Kasner would measure the numbers off in feet with a yardstick along the chalk tray of the blackboard. In this way he would progress along the chalk tray until the final answer brought him to the other end of the tray. He would then say, "The treasure is buried here." And lo and behold, in a waste basket found standing at the end of the chalk tray, he would unearth a bag of Tootsie Rolls, which he would pass out among the pupils of the class.

Naturally before the class had assembled, Professor Kasner had properly positioned the waste basket with its hidden loot.

99° *An interesting judgment.* Gauss thought so highly of Ferdinand Gotthold Max Eisenstein (1823–1852), known today chiefly for the Eisenstein irreducibility criterion, that he is reputed to have claimed that there were three epoch-making mathematicians: Archimedes, Newton, and Eisenstein.

100° *Athletic mathematician.* Harold Bohr (1887–1951), the younger brother of the better-known physicist, Niels Bohr (1885–1962), in addition to being a noted mathematician was also a famous soccer player. He played on the Danish national soccer team. When he completed his Ph.D. in mathematics, pictures of him in the newspapers showed him holding a soccer ball.

ILLUSTRATION FOR 100°

101° *An absentminded mathematician.* Alfred Errera (1886–1960) was a very wealthy man. He once gave an elaborate and lavish dinner party to honor his friend Paul Lévy (1886–1971), who was noted for his profound absentmindedness. The following day Errera chanced to meet Lévy and he told his friend that he had experienced great pleasure the previous evening. Lévy, in all innocence, asked, "Oh, where were you last evening?"

102° *Re the new math.* In imitation of the Roman statesman and general, Marcus Porcius Cato, who used to close all his speeches with "And by the way, Carthage should be destroyed," Oskar Perron (1880–1975) closed a report of the Academic Senate with the words "And by the way, the so-called new math should be destroyed."

103° *Imbibers.*   There was a group of mathematicians, including Shohat, Tamarkin, and Uspensky, who often drank together. On one occasion, when they were well into their cups, they decided to call up an absent member who was the director of a large observatory. It was in the middle of the night and they were informed that the director was asleep. On their insistence he was awakened and brought to the phone. The group then asked him, "What do you feed the great bear?"

104° *Drawing lots.*   On another occasion, the same group of mathematicians mentioned in Item 103° were again well into their cups when they decided to draw lots as to which of them would die first, second, and so on. It happened that the one who drew the lot to die first did die first, even though he was the youngest of the group. Some years later when Tamarkin learned that Shohat had just died, he was very surprised, because the death was out of order.

105° *Thomas Mann's novel.*   Thomas Mann, the German-born American author and Nobel Prize winner, married the daughter of the cultured and musical mathematician A. Pringsheim (1850–1941). Mann later wrote a novel, *Wälsungenblut,* about a girl who had a twin brother for a lover, the idea for the novel being suggested by Siegmund and Sieglinde, the brother and sister of Wagnerian opera. But Mrs. Mann had a twin brother, and the Pringsheim family feared readers would think the novel referred to her and her brother. So Pringsheim, who was a multimillionaire, bought up the entire edition of the novel and had Mann promise not to have the work reprinted. Later, of course, the book was reprinted.

106° *The bookkeeper and the surveyor.*   In 1806 a paper was published entitled "Essai sur une manière de représenter les quantités imaginaires dans les constructions géométriques," in which the now familiar and fruitful association of the complex numbers with the points of a plane was described for apparently the first time. The author of the paper was Jean Robert Argand (1768–

1822), a bookkeeper born in Geneva, Switzerland. It was natural that the plane of complex numbers came to be called, by many, the *Argand plane*.

Many years later, in 1897, an antiquary, poring through some old Danish archives, unearthed a paper entitled "Om Directionens analytiske Betregning" (On the Analytical Representation of Vectors), in which the association of the complex numbers with the points of a plane was clearly set forth. The paper had been presented to the Royal Danish Academy of Sciences in 1797 and then published in that academy's *Transactions* in 1799. It was written by Casper Wessel (1745–1818), a surveyor born in Josrud, Norway. The Wessel paper was republished in a prominent European journal on the hundredth anniversary of its first appearance, and the mathematicians of central Europe thus learned that Wessel has anticipated Argand by some nine years.

Later historical research revealed that the representation had been known to Karl Friedrich Gauss (1777–1855) perhaps even before Wessel; the idea is certainly found in Gauss's doctoral dissertation of 1799. So today many mathematicians refer to the plane of complex numbers as the *Gauss plane*.

One is reminded of the very similar story about János Bolyai, Lobachevsky, and Gauss in connection with the discovery of the first non-Euclidean geometry.

107° *An unusual bid to fame.* Benjamin Peirce, (1809–1880), who was very influential in early American mathematics, was associated with Harvard College for over fifty years, first as a student and then as a professor. He was one of the first American mathematicians to indulge in significant research, to publish valuable papers, and to instill the idea of research among his students. The result is that his fame today seems to rest chiefly on the fact that it was he who first caused the powers-that-be to recognize that mathematical research is one of the reasons for the existence of departments of mathematics in America.

108° *A thought stimulator.* As we have pointed out elsewhere (see Item 360° in *Mathematical Circles Squared*), many scholars find that concentration and the thinking process are stimulated by regular and rhythmic pacing—the "legs are the wheels of thought." Henri Poincaré (1854–1912) said he did his best thinking while restlessly pacing about. There is a story that a little eight-year-old boy once asked Poincaré how to "think mathematically." "You look for an inclined sandy path," Poincaré replied. "You walk up, you walk down, then up again, and down, up and down. Mathematical thinking is generated by the friction of the soles of the feet."

SOME LITERARY SNIPS AND BITS

WRITERS in the purely literary field sometimes parenthetically comment upon, or allude to, the area of mathematics. These remarks range from penetrating insights that delight the mathematician to misunderstood conceptions that amuse him. We have quoted a number of such comments and allusions in the earlier *Circle* books. Here we offer a handful of further examples. As is perhaps to be expected, barring science fiction, mathematical asides appear to occur more frequently in detective fiction than in most other forms of imaginative writing.

109° *Imitation.*   He went through all the details slowly and surely as a mathematician sets up and solves an equation.
—ISAK DINESEN
"The Young Man with the Carnation," in *Winter's Tales.*

110° *A function of time.*   The opening three sentences of O. Henry's famous short story "The Gift of the Magi" are: "One dollar and eighty-seven cents. That was all. And sixty cents of it was in pennies." [Explain.]

111° *A logical conclusion.*   In short, as the Marshall town humorist explained in the columns of *Advance*, "the proposition that the Manton house is badly haunted is the only logical conclusion from the premises."
—AMBROSE BIERCE
The Middle Toe of the Right Foot.

112° *The pen is mightier than the surd.* Don't talk to me of your Archimedes' lever. He was an absentminded person with a mathematical imagination. Mathematics commands all my respect, but I have no use for engines. Give me the right word and the right accent and I will move the world.

—JOSEPH CONRAD
In the preface to *A Personal Record.*

113° *An upper bound.* There was room in [the office] for two chairs, a desk, and a filing cabinet. As for people, any number could enter, providing it was not more than three and they liked each other.

—WILLIAM BANKIER
"The Missing Collectormaniac,"
Alfred Hitchcock Mystery Magazine, Apr. 1981.

114° *One of the greatest detectives of all time.* That amateur sleuth, Average Jones, in Samuel Hopkins Adams's short story "The One Best Bet," saves the life of Governor Arthur by locating, just in time, the position of the would-be assassin's gun. The location of the weapon was accomplished mathematically with the aid of a diagram on the back of an envelope and used triangulation based upon measurements of a certain deflected bullet. At the end of the story, the grateful governor congratulates Average Jones on "having worked out a remarkable and original problem."

" 'Original?' said Average Jones, eyeing the diagram on the envelope's back, with his quaint smile. 'Why, Governor, you're giving me too much credit. It was worked out by one of the greatest detectives of all time, some two thousand years ago. His name was Euclid.' "

115° *No room for argument.* I'm sorry to say that the subject I most disliked was mathematics. I have thought about it. I think the reason was that mathematics leaves no room for argument. If you made a mistake, that was all there was to it.

—MALCOLM X
Mascot.

116° *The complex problem of woman.*   In short, woman was a problem which, since Mr. Brooke's mind felt blank before it, could be hardly less complicated than the revolution of an irregular solid.

—GEORGE ELIOT
Middlemarch.

117° *Solving problems.*   On the other side of the fudge-colored wall the circular saw in the woodworking shop whined and gasped and then whined again; it bit off pieces of wood with a rising, somehow terrorized inflection—*bzzzzzup!* He solved ten problems in trigonometry. His mind cut neatly through their knots and separated them, neat stiff squares of answer, one by one from the long but finite plank of problems that connected Plane Geometry with Solid.

—JOHN UPDIKE
"A Sense of Shelter," *Pigeon Feathers and Other Stories.*

118° *An apt phrase.*   The American poet Conrad Aiken (born in Savannah, Georgia, in 1889), in his poem "At a Concert of Music," uses the apt phrase: "the music's pure algebra of enchantment."

119° *The value of twice one.*   Through all this ordeal his root horror had been isolation, and there are no words to express the abyss between isolation and having one ally. It may be conceded to the mathematician that four is twice two. But two is not twice one; two is two thousand times one. That is why, in spite of a hundred disadvantages, the world will always return to monogamy.

—GILBERT KEITH CHESTERTON
The Man Who Was Thursday.

120° *Squares.*   His face was somewhat square, his jaw was square, his shoulders were square, even his jacket was square. Indeed, in the wild school of caricature then current, Mr. Max

48

Beerbohm had represented him as a proposition in the fourth book of Euclid.

—GILBERT KEITH CHESTERTON
"The Man in the Passage," *The Wisdom of Father Brown.*

[Propositions 6, 7, 8, and 9 of Book IV of Euclid's *Elements* concern themselves with the construction of squares.]

121° *Probability.* Superintendent Leeyes was unsympathetic. "You've had over twenty-four hours already Sloan. The probability that a crime will be solved diminishes in direct proportion to the time that elapses afterward, not as you might think in an inverse ratio."

"No, sir." Was that from "Mathematics for the Average Adult" or "Logic"?

—CATHERINE AIRD
The Religious Body.

122° *A mathematical law?* He knew there were mathematical formulae where time and distance were locked together. No one had yet put a name or symbol to the ratio, but time and crime were inextricably interwoven, too. The importance of justice seemed to vary in inverse proportion to the distance of the crime from the time it was brought to book.

—CATHERINE AIRD
Harm's Way.

123° *Common sense.* "And, as I say, I can do plain arithmetic. If Jones has eight bananas and Brown takes ten away from him, how many will Jones have left? That's the kind of sum people like to pretend has a simple answer. They won't admit, first, that Brown can't do it—and second, that there won't be an answer in plus bananas!"

"They prefer the answer to be a conjuring trick?"

"Exactly. Politicians are just as bad. But I've always held out for plain common sense. You can't beat it, you know, in the end."

—AGATHA CHRISTIE
An Overdose of Death.

49

ILLUSTRATION FOR 123°

124° *Reductio ad absurdum.*   "Therefore it seems impossible that it was anybody—which is absurd!"
"As our old friend Euclid says," murmured Poirot.
—AGATHA CHRISTIE
Murder on the Orient Express.

125° *The perils of probability.*   "I think you're begging the question," said Haydock, "and I can see looming ahead one of those terrible exercises in probability where six men have white hats and six men have black hats and you have to work it out by mathematics how likely it is that the hats will get mixed up and in what proportion. If you start thinking about things like that, you would go round the bend. Let me assure you of that!"
—AGATHA CHRISTIE
The Mirror Crack'd.

126° *Enthralling arithmetic.*   I continued to do arithmetic with my father, passing proudly through fractions to decimals. I eventually arrived at the point where so many cows ate so much grass, and tanks filled with water in so many hours—I found it quite enthralling.
—AGATHA CHRISTIE
An Autobiography.

127° *Reductio ad absurdum again.*   I also remember a tiresome cousin, an adult, insisting teasingly that my blue beads were green and my green ones were blue. My feelings were those of Euclid: "which is absurd," but politely I did not contradict her.
—AGATHA CHRISTIE
An Autobiography.

128° *Euclid restores sanity.*   There is an interesting passage in Lord Dunsany's *The Ghosts* where a man in a deranged state plans to murder his brother. At the crucial moment, just before carrying out his plan, his mind turns to the proposition in Euclid's *Elements* where vertical angles are proved equal. Carefully going over the proof returns him to the world of logic and reason; he is yanked back from the edge and is restored to sanity.

SHERLOCKIANA

BECAUSE of special affection for Sir Arthur Conan Doyle's great detective, Sherlock Holmes, we here record, in a section of its own, allusions to mathematics found scattered throughout the Holmes saga. References to logic are omitted, being too numerous to include.

ILLUSTRATION FOR 129°–134°

129° *Similar triangles.*   In the short story "The Musgrave Ritual," Holmes locates the would-be position, at a given time of day, of the tip of the shadow of a long-ago felled sixty-four foot elm tree. Setting up two lengths of a fishing rod, which came to just six feet, at the spot of the former elm tree, Holmes found the shadow cast by the rod to be nine feet. Therefore a tree of sixty-four feet would throw a shadow of ninety-six feet, along the line of the fishing pole's shadow.

130° *Conclusions of a trained observer.*   In Chapter II of *A Study in Scarlet,* we read that Holmes once wrote an article entitled "The Book of Life," in which he claimed that the conclusions of one trained to observation and analysis would be "as infallible as so many propositions of Euclid. So startling would his results appear to the uninitiated that until they learned the processes by which he had arrived at them they would well consider him a necromancer."

131° *An elopement in Euclid's fifth proposition.*   In Chapter II of *The Sign of Four,* acknowledging that he had perused the account of his solution of the Jefferson Hope case as narrated by Watson in *A Study in Scarlet,* Holmes deflates his friend by re-marking, "I glanced over it. Honestly, I cannot congratulate you upon it. Detection is, or ought to be, an exact science and should be treated in the same cold and unemotional manner. You have attempted to tinge it with romanticism, which produces the same effect as if you worked a love-story or an elopement into the fifth proposition of Euclid."

[The fifth proposition of Euclid, which proves that the base angles of an isosceles triangle are equal, became known as the *pons assinorum,* or asses' bridge, a reference to the bridgelike ap-pearance of the figure accompanying the proposition and the fact that many beginners experience difficulty in "getting over" it.]

132° *The rule of three.*   There is another mathematical ref-erence in *The Sign of Four,* this one in Chapter VI. A small barefoot Andaman Islander named Tonga inadvertently stepped into a puddle of creosote, thus rendering it an easy task to track him down. Holmes comments, "I know a dog that would follow that scent to the world's end. If a pack can track a trailed herring across a shire, how far can a specially trained hound follow so pungent a smell as this? It sounds like a sum in the rule of three."

[The rule of three states the method of finding the fourth term x of a proportion $a : b = c : x$, where a, b, c are known.]

133° *Professor Moriarty and the binomial theorem.*   In de-scribing, in "The Final Problem," his great arch enemy Professor

James Moriarty, Holmes says, "His career has been an extraordinary one. He is a man of good birth and excellent education,
endowed with a phenomenal mathematical faculty. At the age of
twenty-one he wrote a treatise on the binomial theorem, which
has had a European vogue. On the strength of it he won the
mathematical chair at one of our small universities, and had, to
all appearance, a most brilliant career before him."

[Many eminent names in mathematics, Isaac Newton among
them, have become associated with the binomial theorem.]

134° *Professor Moriarty and "The Dynamics of an Asteroid."* In reply to some harsh words uttered by Watson about
Professor Moriarty, Holmes, in *The Valley of Fear,* comments, "But
so aloof is he from general suspicion, so immune from criticism,
so admirable in his management of self-effacement, that for those
very words that you have uttered he could hale you to a court
and emerge with your year's pension as a solatium for his wounded
character. Is he not the celebrated author of *The Dynamics of an
Asteroid,* a book which ascends to such rarified heights of pure
mathematics that it is said there was no man in the scientific press
capable of criticizing it? Foul-mouthed doctor and slandered professor—such would be your respective roles! That's genius, Watson. But if I am spared by lesser men, my day will surely come."

POETRY, RHYMES, AND JINGLES

It would not be difficult to construct a fair-sized volume devoted
only to poetry, rhymes, and jingles that allude to mathematics and
its concepts. Here is a batch in addition to the examples appearing
elsewhere in the *Circle* books.

135° *Lines found written in a college mathematics text.*
    If there should be another flood,
       For refuge hither fly;
    Though all the world would be submerged,
       This book would still be dry.

136° *Relativity.*

> In every way in which we live,
> Our values are comparative,
> Observe the snail who with a sigh,
> Says: "See those turtles whizzing by."

> —Unknown

137° *Elliptic integrals.*

> There was a mathematician named Nick,
> For whom integration was a kick,
>> But an elliptic arc
>> Finally left its mark,
> It was something he could not lick.

> —Unknown
> Heard at a mathematics meeting.

138° *Was the bridge refereed?*

> Hamilton won acclaim (so people tell)
> Carving equations on a bridge; oh, well,
> If you or I would try to do that now,
> I'd hate to contemplate the awful row,
> Surely the cops would come and run us in;
> Putting graffiti on a bridge is sin.
> Were it allowed, it wouldn't even pay—
> Call it a publication? There's no way.

> —Anonymous
> *Two-Year College Mathematics Journal,* Jan. 1979.

139° *Re pure mathematics.*

> If it's pure, it's sterilized.
> If it's sterile, it has no life in it.
> If it has no life in it, it's dead.
> If it's dead, it's putrified.
> If it's putrified, it stinks.

> —A former student

54

140° *The law of the syllogism.*

    If there be righteousness in the heart,
      There will be beauty in the charac-
    ter.
    If there be beauty in the character,
      There will be harmony in the home.

    If there be harmony in the home,
      There will be order in the nation.
    If there be order in the nation,
      There will be peace in the world.

                        —Chinese Proverb

141° *A valentine jingle.*

    You are the fairest of your Sex,
      Let me be your hero.
    I love you like one over "x"
      As "x" approaches zero.

                            —Anonymous
         Contributed by Michael A. Stueben.

142° *Shanks's famous error.*

    Seven hundred seven, Shanks did state,
    Digits of π he would calculate.
      And none can deny
      It was a good try.
    But he had erred in five twenty eight.

                       —NICHOLAS J. ROSE
      Rome Press 1985 Mathematical Calendar.

143° *The equality of null sets.*

    The man in the wilderness asked of me
      How many strawberries grew in the sea.
    I answered him as I thought good,
    As many as red herring grow in the wood.

                          —Anonymous

144° *A limerick.*

> Chicago's mathematical forces,
> In spite of their numerous resources,
>   Always adorn
>   With a lemma of Zorn
> At least 90% of their courses.

> —Unknown

145° *Calculating machines.*

> I'm sick and tired of this machine
> I wish that they would sell it.
> It never does just what I want,
> But only what I tell it.

> —Unknown

146° *Paradox.*

> How quaint the ways of paradox—
> At common sense she gaily mocks.

> —W. S. GILBERT

147° *Getting Down to the Nitty-Gritty.*

> Biologists analyze the cell
> Chemists manipulate the molecule
> Physicists dissect the atom
> Mathematicians idealize the point

> —RAY BOBO

148° *The Higher Pantheon in a Nutshell.*

Doubt is faith in the main; but faith, on the whole is doubt;
We cannot believe by proof; but could we believe without?

One and two are not one; but one and nothing is two;
Truth can hardly be false, if falsehood cannot be true.

> —ALGERNON CHARLES SWINBURNE

149° *Newton and the apple.*

> When Newton saw an apple fall, he found
> A mode of proving that the earth turn'd round—

In a most natural whirl, called gravitation;
And thus is the sole mortal who could grapple
Since Adam, with a fall or with an apple.
                    —LORD GEORGE GORDON BYRON

150° *A clerihew.*
            George Boole
            was nobody's fool:
            but never forget—
    his mathematical legacy is the empty set.
                    —ROBIN HARTE
        American Mathematical Monthly, Nov. 1985

151° *An inversion transformation.*
            There was a young lady of Niger
            Who smiled as she rode on a Tiger;
                They came back from the ride
                With the lady inside,
            And the smile on the face of the Tiger.
                    —Anonymous

152° *The set of all sets which belong to themselves.*
            Bertrand Russell was most mortified,
            When a box was washed up by the tide,
                For he said with regret,
                "Why, the set of all sets
            Which belong to themselves is inside."
                    —PAUL RITGER
        Rome Press 1986 Mathematical Calendar.

153° *Achilles and the tortoise.*
            When Zeno was still a young man
            Impressed with the way turtles ran,
                He challenged Achilles
                And some say that still he's
            Not certain which one's in the van.
                    —PAUL RITGER
        Rome Press 1985 Mathematical Calendar.

154° *The integers.*

> Of the integers, so we are told,
> No matter how many are bought or sold,
> Or how many you give away or lend,
> Of the rest there is no end.

—Unknown

155° *Song of the Screw.*

> A moving form or rigid mass,
>   Under whate'er conditions
> Along successive screws must pass
>   Between each two positions.
> It turns around and slides along—
> This is the burden of my song.
>
> The pitch of screw, if multiplied
>   By angle of rotation,
> Will give the distance it must glide
>   In motion of translation.
> Infinite pitch means pure translation,
> And zero pitch means pure rotation.
>
> Two motions on two given screws,
>   With amplitudes at pleasure,
> Into a third screw-motion fuse;
>   Whose amplitude we measure
> By parallelogram construction
> (A very obvious deduction.)
>
> Its axis cuts the nodal line
>   Which to both screws is normal,
> And generates a form devine,
>   Whose name, in language formal,
> Is "surface-ruled of third degree."
> Cylindroid is the name for me.
>
> Rotation round a given line
>   Is like a force along.

If to say couple you incline,
    You're clearly in the wrong;—
'Tis obvious, upon reflection,
A line is not a mere direction.

So couples with translations too
    In all respects agree;
And thus there centres in the screw
    A wondrous harmony
Of Kinematics and of Statics,—
The sweetest thing in mathematics.

The forces on one given screw,
    With motion on a second,
In general some work will do,
    Whose magnitude is reckoned
By angle, force, and what we call
The coefficient virtual.

Rotation now to force convert,
    And force into rotation;
Unchanged the work, we can assert,
    In spite of transformation.
And if two screws no work can claim,
Reciprocal will be their name.

Five numbers will a screw define,
    A screwing motion, six;
For four will give the axial line,
    One more the pitch will fix;
And hence we always can contrive
One screw reciprocal to five.

Screws—two, three, four or five, combined
    (No question here of six),
Yield other screws which are confined
    Within one screw complex.
Thus we obtain the clearest notion
Of freedom and constraint of motion.

In complex III, three several screws
   At every point you find,
Or if you one direction choose,
   One screw is to your mind;
And complexes of order III
Their own reciprocals may be.

In IV, wherever you arrive,
   You find of screws a cone,
On every line in Complex V
   There is precisely one;
At each point of this complex rich,
A plane of screws have a given pitch.

But time would fail me to discourse
   Of Order and Degree;
Of Impulse, Energy, and Force,
   And Reciprocity.
All these and more, for motions small,
Have been discussed by Dr. Ball.

—Anonymous

COMPUTERS AND CALCULATORS

MANY stories and jokes about modern electronic computers and calculators tend to belittle and ridicule these marvels. This attitude probably reflects the distrust and fear that frequently accompany ignorance and lack of understanding.

156° *The ever-recurring excuse.*   We like to poke fun at the old-time bookkeeper who sat on a high stool and recorded his figures slowly and meticulously with a quill. But you never heard the excuse, "Our computer is down."

157° *An important date.*   In September, 1971, the first pocket calculator was offered for sale in the consumer market; Bowmar Instrument Corp. of Fort Wayne, Indiana introduced a model that measured 3-by-5 inches and sold for $249.

ILLUSTRATION FOR 156°

Bowmar soon had plenty of company. Within a year and a half nearly a dozen firms were selling calculators in the stores. With fierce competition, the prices plummeted. By Christmas 1972, the lowest-priced calculators fell under the $100 mark, opening the discount store market. A year later, the price was below $50, and now models are available for less than $10. By 1974, annual sales topped $10,000,000.

As calculators got cheaper, new battery designs also allowed designers to slim them down to roughly the thickness of a credit card. Today, calculators are perhaps the third or fourth largest selling consumer products, with annual retail sales of $500,000,000 to $700,000,000.—*Bangor Daily News,* Sept. 19–20, 1981.

158° *Calculators in the classroom.*   In 1979 the National Council of Teachers of Mathematics conducted a survey of teachers of education and education leaders and found they were quite conservative about the use of calculators in the classroom. James D. Gates, executive director of the Council, says there was general support for using calculators to check homework answers, but not much else.

"There is a feeling that kids will use it as a crutch," Gates said, "Teachers are still very cautious. There is not enough research to be sure it is not going to damage mathematical skills."

Not so, says Marilyn N. Suydam, executive director of the federally funded Calculator Information Center at Ohio State University. According to Miss Suydam, nearly 100 studies have shown that the use of calculators either doesn't hurt mathematics achievement or actually improves achievement.

"Computation is not the problem," she said. "Most kids have a mastery of that within two or three years. The problem that kids have is problem solving—that is, applying their computational skills to solving problems."

As a practical matter, it is usually left up to individual teachers to decide whether they will use calculators in their classes, Miss Suydam said.—*Bangor Daily News,* Sept. 19–20, 1981.

159° *A recent (1986) blurb.*   Syracuse University has begun a teacher education class taught entirely by computer.

160° *An interesting little-known fact.*   It is well known that Charles Babbage devoted 37 years, a large part of his personal fortune, and several government grants to his Analytical Engine. But it is little known that, for a time, Lord Byron's only legitimate daughter, Augusta Ada Byron, the Countess of Lovelace, wrote programs for it—the better to play the horses, it was rumored.

161° *A computing contest.* In a contest between a calculator and a fifteenth-century Chinese abacus in Melbourne, Australia, the abacus won nine times out of ten.

162° *The real danger.* The real danger of our technical age is not so much that machines will begin to think like man, but that man will begin to think like machines.—SYDNEY J. HARRIS

163° *A cautionary tale.* A hydrodynamicist was reading a research paper translated from the Russian and was puzzled by references to a "water sheep." It transpired that the paper had been translated by a computer; the phrase in question should have been "hydraulic ram."

164° *The word computer.* A word computer was asked to paraphrase the sentence, "He was bent on seeing her." The computer wrote out, "The sight of her doubled him up."

165° *A special program.* Since the computer's storage space was becoming cluttered with little-used programs, the systems programmer wrote a program to seek out rarely used programs and delete them. Put into operation, the program dutifully looked for the least-used program—and promptly erased itself!
—THOMAS R. DAVIS
Two-Year College Mathematics Journal, Jan. 1979.

166° *Boy to father.* "It's okay for you to say arithmetic is easy, you figure in your head—I have to use a computer!"
—GEORGE LEVINE
National Enquirer.

167° *An important point.* Man has made some machines that can answer questions, provided the facts are previously stored

in them, but he will never be able to make a machine that will ask
questions. . . . The ability to ask the right questions is more than
half the battle of finding the right answers.—TOM WATSON, JR.

168° *Chisanbop.*   Chisanbop, which means "finger calculation method" in Korea, was conceived in the late 1950s by Sung
Jin Pai, a noted Korean mathematician. His son, Hang Young Pai,
began teaching the method, shortly after his arrival in the United
States in 1976, to students of the Korean-American School in New
York. He quietly taught Chisanbop and drew little attention. Then,
quite by accident, Edwin Lieberthal, a marketing executive, learned
of Pai's work.

During demonstrations on the "Today" and "Tonight" shows,
grade school children accurately added series of four- and five-
digit numbers at electronic calculator speeds, even beating the
time of a college math professor who used a pocket calculator.

The basis of the Chisanbop method lies in assigning numbers
to the fingers and thumbs of the two hands so that you can easily
"hold" any number from 1 through 99, thereby keeping a running
tally in an arithmetic operation.

Part (a) of the illustration shows the finger values. By pressing
down a combination of fingers and thumbs against a desk or other
hard surface, you can represent any number from 1 through 99.

Part (b) of the illustration provides a simple example of the
method: the addition of 18 and 16. First, represent the 18 by
pressing down the left index finger and the right thumb, index,
middle, and ring fingers. To add 16, think of it as 10 plus 5 plus
1. Add the 10 by pressing down the middle finger on your left
hand, keeping down the fingers already pressed. To add the 5,
you have to make an exchange between the hands. Lift your right
thumb, which subtracts 5; then, put down your left ring finger,
which adds 10, for a net addition of 5. To add the 1, press down
your right little finger. Read off the answer from the way your
fingers and thumb are now pressed.

—Adapted from an article by RUTH FOSTER
Family Week, July 8, 1978.

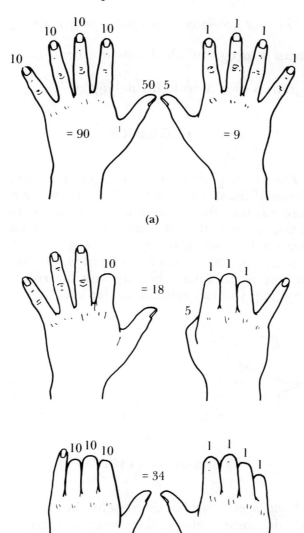

(a)

(b)

ILLUSTRATION FOR 168°

65

169° *The first two laws of computer programming.*

1. Never trust a program which has not been thoroughly debugged.
2. No program is ever thoroughly debugged.

ALGEBRA

170° *The rules of algebra.*   An algebra teacher, trying to convince some of his students who felt restricted by many of the rules of algebra, told the story of a kite. It seems that the kite wanted its tail cut in half so it could be freer to roam the skies. Its tail was cut in half, and it swooped to the ground. "Now boys," admonished the algebra teacher, "the tail wasn't hindering the kite, it was helping it. That's the way with the rules of algebra; actually they help you rather than hinder you."

ILLUSTRATION FOR 170°

171° *Symbolism.*   Symbolism is useful because it makes things difficult. Obviousness is always the enemy of correctness. Hence we must invent a new and difficult symbolism in which nothing is obvious. The whole of arithmetic and algebra has been shown to require three indefinable notions and five indemonstrable propositions.

—BERTRAND RUSSELL
International Monthly, 1901.

172° *A fable about Ph.D. oral exams.* On one particular examining board sat an older algebraist known for his particularly unpleasant questions. He asked a graduate student the following question: "Give three essentially different equations which represent a straight line." The student thought for a time and said that he could only come up with $y = mx + b$. With that the professor quipped, "Well, there is $y = \log x$ on log paper and $y = \log \log x$ on log log paper." At this point one of the younger professors interjected, "Yeah, and there is also $y = f(x)$ on f-paper!"

—ROBERT J. KLEINHENZ
Two-Year College Mathematics Journal, Jan. 1979.

173° *Another fable about Ph.D. oral exams.* One graduate student was asked to produce an integral domain that was not a UFD. The student answered: "Z extended by $\sqrt{-5}$." The examiner asked him why. The student replied that 6 was not uniquely factorable. "Give us one factorization," asked the board. The student wrote $6 = (1 + \sqrt{-5})(1 - \sqrt{-5})$. When questioned further about another factorization the student was at a loss to produce a different one!

—ROBERT J. KLEINHENZ
Two-Year College Mathematics Journal, Jan. 1979.

174° *Hatred of logarithms.* Everybody knows (or at least knew, before they all had pocket calculators) that with logarithms one can reduce the odius task of multiplying two long numbers to the merely distasteful one of adding two others, and a little hunting through tables. By having many sunny hours of our adolescense filled with such pursuits, some of us have acquired a deep hatred of logarithms.

—N. DAVID MERMIN
"Logarithms!," *American Mathematical Monthly,* Jan. 1980.

175° *What price common logs?* Seen last summer, in the window of a store in Amagansett, Long Island, N.Y., beside a display of Royal Oak Brix Instant Charcoal, a sign reading

NATURAL
LOGS
49¢ each

—ALAN WAYNE

176° *I see.*   Old math teachers never die, they just multiply.

177° *Oh!*   Old accountants never die, they just lose their balance.

178° *Addition and subtraction.*   If you would make a man happy, do not add to his possessions but subtract from the sum of his desires.—SENECA

179° *What's in a name?*

Wayne: Take for example a former patient of mine; let us call him Mr. X.
Schuster: Why?
Wayne: All right, let us call him Mr. Y. Now Mr. Y, formerly Mr. X. . . .

—From a Wayne and Schuster television skit.

180° *Dolbear's law.*   It has been known for some time that the frequency of a cricket's chirps varies with temperature. This thermal variation makes it possible to compute the temperature by counting chirps. The idea was first proposed in 1897 by A. E. Dolbear, a physics professor at Tufts University. In an article titled "The Cricket As a Thermometer," Professor Dolbear gave the following formula, which has become known as Dolbear's Law:

$$T = 50 + (N - 40)/4,$$

where T stands for Fahrenheit temperature and N for the number of chirps per second.

Different species have different chirp rates, and later investigators have suggested formulas for several species. If the species you are studying is the snowy tree cricket, whose chirps are particularly clear and consistent, you may simply count the pulsations per minute, divide by 4, and add 40. Or you may count the number of chirps in 15 seconds and add 40. Scientists have been able to trick female crickets into approaching males of the wrong species by raising the temperature in the cage where the males were kept.

QUADRANT THREE

From impossible geometry
to an induction problem

GEOMETRY

181° *Impossible geometry.* A large merchandise company sells a smoke detector with a circular base 7 inches in diameter. Under the instructions for mounting the detector on a wall we read that it must be placed with "6 inches minimum distance from ceiling to top of detector base and 12 inches maximum from ceiling to bottom of detector base."

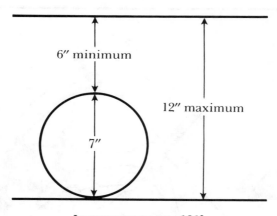

182° *Political geometry.* Former Secretary of State Henry A. Kissinger once remarked that China and the United States "can pursue parallel policies where their interests converge" while remaining in civil disagreement in other areas.

183° *Advice to a geometry student.* There's a mighty big difference between good, sound reasons and reasons that sound good.

184° *The Great Pyramid and pi.* One of the puzzling features of the Great Pyramid of Egypt is the noted fact that the ratio of twice a side of the square base to the pyramid's height yields a surprisingly accurate approximation of the number pi, and it has been conjectured that the Egyptians purposely incorporated this ratio in the construction of the pyramid. Herodotus, on the other hand, has stated that the pyramid was built so that the area of each lateral face would equal the area of a square with side equal to the pyramid's height. Show that, if Herodotus is right, the concerned ratio is automatically a remarkable approximation of pi.

ILLUSTRATION FOR 185°

185° *An ardent pyramidologist.* The famous archeologist Sir Flinders Petrie, who conducted some very precise measurement of the Great Pyramid of Egypt, has reported that he once caught a pyramidologist stealthily filing down a projecting stone to make it agree with one of his theories.

186° *Definition of a circle.* Someone said that Mark Twain described a circle as a straight line with a hole in it.

187° *Why we study mathematics.* Like Euclid's geometry student of long ago, a mathematics student inquired of his professor, "What's the good of this stuff anyway? Where does it get you?"

The professor quietly replied, "The good of it is that you climb mountains."

"Climb mountains," retorted the student, unimpressed. "And what's the use of doing that?"

"To see other mountains to climb," was the reply. "When you are no longer interested in climbing mountains to see other mountains to climb, life is over."

188° *The reason for self-imposed limitations.* A geometry student once asked his teacher why, in geometry, students must limit constructions to the use of only straightedge and compasses, or to just the straightedge or the compasses alone?

The teacher replied that he was reminded of an elderly lady who saw how often the ball hit the net during a tennis match. Exasperated, she declared, "Why don't they take down the net?"

Some folks cannot comprehend the value of obstacles and opposition. They never realize the satisfaction and exhilaration experienced by winning against odds.

189° *Books and the fourth dimension.* Books are quiet; they do not suddenly stop functioning, nor are they subject to wavy lines and snowstorms. There is no pause for commercials. They are small and compact for convenience in handling. From a purely materialistic standpoint they are three dimensional, having length, breadth, and thickness, but they live indefinitely in the fourth dimension of time.

190° *Concerning a geometry book.*   "It's a mighty good geometry book," said a mathematics professor, speaking of one of his own publications, to a student. "Have you read it? What do you think of it?"

"There is only one thing to be said in its favor," replied the student. "A friend of mine carried it through the war in his breast pocket. A bullet ricocheted against his ribs, but the book saved him. The bullet was unable to get beyond the problems at the end of the first section."

191° *A matter of distance.*   I don't believe in special providences. When a mule kicks a man and knocks him anywhere from eight to twenty feet, I don't lay it on the Lord; I say to myself, "That man got a little too near the mule."

192° *A triple-barreled pun.*   Allegedly the only triple pun in English: a widow, giving her cattle ranch to her boys, named it Focus—where the sons raise meat.

193° *Time often tells.*   The young and restless Alexander the Great was tutored by Menaechmus, and he found the study of mathematics too slow and too boring. Of what possible use, he must have thought in his youthful hurry to conquer the world, could Menaechmus's conic sections ever play on the field of battle? [The modern science of gunnery is based upon a knowledge of the conic sections.]

194° *The reason why.*   A very old tale serves well to illustrate the manner in which geometry plunges below the surface. In days before Euclid the plodding student encountered the unpleasant gibe that in proving that two sides of a triangle were together greater than the third, he stated no more than every donkey knew. The fact was well known, so it was derisively observed, to every donkey who walks straight to a bunch of hay. Nevertheless the student had a sufficiently good retort: "Yes, but the donkey does not know the reason why!"

—W.B. FRANKLAND
The Story of Euclid. Hodder and Stoughton, 1901.

195° *Progress.*   Whereas at the outset geometry is reported to have concerned herself with the measurement of muddy land, she now handles celestial as well as terrestial problems: she has extended her domain to the furthest bounds of space.

—W.B. FRANKLAND
The Story of Euclid. Hodder and Stoughton, 1901.

196° *A proportion.*   Proclus is to Euclid as Boswell is to Johnson.

197° *A mermaid.*   Bob Rosenbaum commented on the lame ending of Saccheri's *Euclid Freed of Every Flaw* by saying, "It reminds me of a mermaid; the torso is beautiful, but it has a fishy end."

198° *A measurement of distance.*   Different cultures have evolved different ways of measuring distance. For example, some American Indians measured a great distance by the number of days it would take to make the journey. In Laos, distance is measured in terms of how long it takes to cook rice.

199° *An old definition.*   *Geometer:* a species of caterpillar.

—Old Dictionary

ILLUSTRATION FOR 199°

200° *A crackpot in academia.*   One isn't surprised when an unschooled person diligently seeks to trisect the general angle, square the circle, duplicate the cube, prove the parallel postulate, or invent a perpetual motion machine or an antigravity screen. Nor is the surprise great when an elected politician (and there have been a goodly number) attempts these things. But one certainly must register surprise when the president of an eminent university turns his hand to such matters.

In 1931 the Very Reverend Jeremiah J. Callahan, the president of Duquesne University in Pittsburgh, launched a "proof" that the Einstein relativity theory is false and sheer nonsense. The idea of how to achieve this "proof" occurred to Father Callahan one day when riding the New York City subway. Since relativity theory is developed against a backdrop of non-Euclidean geometry, all one must do to discredit relativity theory, reasoned Father Callahan, is to show the logical inconsistency of non-Euclidean geometry, and this will be accomplished by deriving the parallel postulate from Euclid's other postulates. It was in 1931 that Father Callahan published his derivation of the parallel postulate in a 310-page work titled *Euclid or Einstein,* wherein any capable geometer can easily locate the error in the good Father's proof.

In that same year, 1931, Duquesne University published a pamphlet written by Reverend Callahan in which a general angle is purported to be trisected with compasses and straightedge, and an announcement was made that the author was working on the duplication of the cube and the squaring of the circle. Father Callahan retired as president of Duquesne University in 1940, at the age of 62.

201° *Trisectors.*   Angle trisectors are rarely overcome by argument. Since their work is not founded on reason, it cannot be destroyed by logic.

202° *Advice.*   It is never wise to argue with an angle trisector; onlookers cannot discern which is the fool.

203° *An Andy Griffith story.*   Andy Griffith tells a story about a West Virginia hillbilly who managed to scrape together enough money to send his son to college. When the boy came home for the Christmas holidays, the father asked him if he had learned anything at college.

"Oh, yes," replied the boy.

"That's fine," said the father. "Give me an example of what you learned."

"Well," said the boy, "I learned πr^2."

"Pie are square?" queried the amazed father. "Pie are not square; pie are round. Corn pone are square."

204° *A truism.*   All geometrical reasoning is, in the last result, circular.

—BERTRAND RUSSELL
Foundations of Geometry. Cambridge University Press, 1897.

205° *Another truism.*   Geometry is the art of correct reasoning on incorrect figures.

—GEORGE PÓLYA
How to Solve It. Princeton University Press, 1945.

206° *What is a curve?*   Everyone knows what a curve is, until he has studied enough mathematics to become confused through the countless number of possible exceptions.

—FELIX KLEIN
On Mathematics, edited by Robert Moritz. Dover, 1958.

207° *Perhaps.*   Perhaps analytic geometry can be regarded as the royal road to geometry that Euclid thought did not exist. [See Item 67° of *In Mathematical Circles.*]

208° *A geodesic.*   The highway of fear is the shortest route to defeat.

209° *Square people.* We don't need any more well-rounded people. We have too many now. A well-rounded person is like a ball; he rolls in the first direction he is pushed. We need more square people who won't roll when they are pushed.

—Eugene Wilson

210° *Why the world is round.* The world is round so that friendship may circle it.

NUMBERS

211° *Children in Germany.* I came back from Europe thinking German families were very large. Many times when I would ask Germans if they had any children they would reply, "Nein."

212° *Martinis in Germany.* An American businessman went into a bar in West Berlin and ordered a dry martini. When the bartender appeared to be confused, the American repeated, "dry martini." The bartender shrugged and fixed the man three martinis.

213° *On the ski slopes in Austria.* At the winter Olympics in Innsbruck, Austria, the starter of the downhill ski race would start the skiers with the following count: "Ein, zwei, drei, vier." Before his turn on a rather snowy and hazardous day, one American said to another, "We have nothing to fear but *vier* itself."

214° *How big is a billion?* If one spent $1,000 a day since the day Christ was born, one would not yet (1987) have spent a billion dollars. Indeed, far from it. One would still have more than a quarter of a billion dollars left—enough to go on spending $1,000 a day for another 750 years.

215° *Ignorance of the law is no excuse.* There are 2,000,000 laws in force in the United States. If a man could familiarize himself with them at the rate of ten each day, he could qualify to act as a law-abiding citizen in the short space of six thousand years.

216° *The federal budget.*   The federal budget is given in terms of billions of dollars. Since few of us can comprehend the enormous magnitude of a billion, various descriptions have been given to shock us into some sort of realization of its incredible size. One such description states that if a person were to stand next to a large hole in the ground and once every minute, day and night, drop a $20 bill into the hole, it would require ninety-eight years to dispose of a billion dollars.

For other stories about large numbers, see Items 25° through 36° in *Mathematical Circles Revisited.*

217° *Bad manners.*   "You should never mention the number 288 in public."

"Why not?"

"It's two gross."

218° *A man is like a fraction.*   A man is like a fraction whose numerator is what he is and whose denominator is what he thinks of himself. The larger the denominator the smaller the fraction.— LEO TOLSTOY

219° *Phenomenal memory ability.*   In 1979 Hans Eberstark, an interpreter for the United Nations Agency, recited, at the European Nuclear Research Center, the decimal expansion of π to a total of 9,744 places, breaking his previous record of 5,050 places.

Eberstark, who was born in Vienna in 1927, can mentally multiply two 8-digit numbers in a matter of seconds, and two 12-digit numbers within two or three minutes. If you tell him the date of your birth, he'll quickly tell you what day of the week it was. He can memorize a random 30-digit number in a few seconds, then repeat it both forwards and backwards. He speaks sixteen "and a half" languages; the half a language being Swiss German. He says he can memorize a thousand digits in an afternoon and could probably memorize a million digits in three or four years if he devoted all his time to it.

Eberstark's rival in the memorization of π is David Sanker of the United States. The two vie with one another for listing in the *Guinness Book of World Records.*

220° *A rumor.*   The equatorial circumference of our earth is stated in most science books as 24,901.5 miles. The rumor is that the 1.5 part is fictitious and was made up to circumvent the idea that 24,900 is accurate only to the nearest hundred miles.

—MICHAEL A. STUEBEN

221° *Of course.*   The teacher is at the blackboard drilling her pupils in the natural number system. "And what comes before 6?" she asks. "The garbage man," replies Tommy.

222° *Slow progress.*

Son: Dad, will you help me find the least common denominator in this problem?

Dad: Good heavens, son, don't tell me that hasn't been found. They were looking for it when I was a kid.

223° *Aztec use of chocolate as a counting unit.*   The Chocolate Information Council has reported that the Aztec Indians of Central America used chocolate not only as a food but also as a counting unit.

Instead of counting by tens as we do, the Aztec Indians, like the Maya Indians of the Yucatán, used 20 as a number base. Numbers up to 20 were represented by dots, 20 was indicated by a flag, and 20^2 was shown by a fir tree. The next unit, 20^3, was represented by a sack of cocoa beans, since each sack contained approximately 8,000 beans.

224° *Numbering the math courses at Harvard.*   The courses in mechanics at Harvard, back around 1930, happened to be numbered in geometric progression. Math 2 was devoted to elementary kinematics, Math 4 to dynamics, Math 8 to the Hamilton-Jacobi theory, and Math 16 was G. D. Birkhoff's course on relativity. H.

Illustration for 223°

W. Brinkmann once proposed a special mechanics course, to be numbered Math 32—but only in jest.

225° *Test for divisibility by 7.* Ralph P. Boas says that when he was serving on the Editorial Board of *Mathematical Reviews,* he came across a paper in which it was proposed that something be done to furnish a simple way of testing integers for divisibility by 7. You merely express the number to base 8 and check if the sum of the digits is divisible by 7, Boas points out.

226° *On adding columns of figures.* There is a cute story told about the former eminent chemist, Gilbert Newton Lewis (1875–1956), of the University of California at Berkeley. One day at the Faculty Club, while he was chatting with members of the department of education, an argument arose as to the best way to teach the operation of addition of figures—should one add the columns from top to bottom or from bottom to top? "That's easy," interrupted Lewis. "I always add each column from top to bottom and then from bottom to top, and I take the average."

During a seminar a student once made a comment about which Lewis remarked, "That was a very impertinent comment, but it was also very pertinent."—Charles W. Trigg

227° *Hidden laws of number.*   It has been observed by many that there are laws of arithmetic one never encounters in school books. For example, everyone knows that if you add a column of numbers from the top down and then again from the bottom up, the result is always different.

228° *After the death of Vern Hoggatt.*   For those of us who knew him, it is natural to imagine Vern now, off on the Eternal Tangent, with 1 robe, 1 halo, 2 wings, and 3 strings of his harp strummed by the 5 fingers of his right hand. . . . Oh, how we miss him, but we smile as we cry.—DAVE LOGOTHETTI

[Hoggatt was an international expert on the Fibonacci and allied sequences and founding editor of *The Fibonacci Quarterly*.]

229° *A criticism.*   L. E. Dickson, during a discussion period that followed the presentation of a paper at a meeting of the American Mathematical Society, criticized the choice of the paper's topic. "It is a lucky thing," he said, "that newspaper reporters do not attend these meetings. If they did, they would see how little our activities are related to the real needs of society." Fifteen minutes later he presented a paper of his own outlining a proof that every sufficiently large integer can be written as a sum of, not 1140 tenth powers (the best previous result), but 1046 tenth powers.

230° *Pólya's reply.*   Let $E_n = P_n + 1$, where $P_n = p_1p_2 \cdots$ p_n is the product of the first n primes. When asked by a student whether E_n is prime for infinitely many values of n, George Pólya is reported to have replied, "There are many questions which fools can ask that wise men cannot answer." The only values of n for which E_n is known to be prime are $n = 1, 2, 3, 4, 5,$ and 11.

231° *Fortunate numbers.*   In 1980, the anthropologist Reo Fortune, once married to the anthropologist Margaret Mead, conjectured that if Q_n is the smallest prime greater than $E_n = P_n + 1$, where $P_n = p_1p_2 \cdots p_n$ is the product of the first n primes, then the difference $F_n = Q_n - P_n$ is *always* prime. The numbers F_n are now, for obvious reasons, called *fortunate numbers*. Fortune's conjecture has not been resolved. The sequence $\{F_n\}$ of fortunate numbers begins

$$3, 5, 7, 13, 23, 17, 19, 23, 37, 61, 67, 61,$$

$$71, 47, 107, 59, 61, 109, 89, 103, 79,$$

and all of these are prime.

(Let's hope Mr. Fortune never had a daughter.)

232° *Smith numbers.*   Professor Albert Wilansky, of Lehigh University, has become interested in those composite numbers each of which has the sum of its digits equal to the sum of all the digits of all its prime factors. Examples of such numbers are 9985 and 6036, inasmuch as

$$9985 = 5 \times 1997, \quad 9 + 9 + 8 + 5 = 5 + 1 + 9 + 9 + 7$$

and

$$6036 = 2 \times 2 \times 3 \times 503, \quad 6 + 0 + 3 + 6 =$$
$$2 + 2 + 3 + 5 + 0 + 3.$$

It has been shown that there are 47 of these numbers between 0 and 999, 32 between 1000 and 1999, 42 between 2000 and 2999, 28 between 3000 and 3999, 33 between 4000 and 4999, 32 between 5000 and 5999, 32 between 6000 and 6999, 37 between 7000 and 7999, 37 between 8000 and 8999, and 40 between 9000 and 9999. Wilansky wonders whether there are infinitely many numbers of this kind. The largest known number of the sort is due to H. Smith, who is not a mathematician but is Wilansky's brother-in-law. The number is 4937775, and is Smith's telephone number. For this reason, Wilansky has called the numbers in question *Smith numbers.*

233° *A truism.*   Almost all positive integers are greater than 1,000,000,000,000.
> —RICHARD COURANT AND HERBERT ROBBINS
> *What is Mathematics?* Oxford University Press, 1941.

234° *Another truism.*   Round numbers are always false.
> —SAMUEL JOHNSON

235° *A Murphy's Law.*   In precise mathematical terms, 1 + 1 = 2, where = is a symbol meaning seldom if ever.
> —ARTHUR BLOCH
> *Murphy's Law: Book Three.* Price/Stern/Sloan, 1982.

236° *Another Murphy's Law.*   For large values of one, one approaches two, for small values of two.
> —ARTHUR BLOCH
> *Murphy's Law: Book Three.* Price/Stern/Sloan, 1982.

237° *A smart animal.*   In my book *In Mathematical Circles* I gave a number of stories supporting the belief that some animals perhaps possess a degree of number sense. A corroborating story has since reached me. It seems that a female mink that was raised on a farm had a litter of five. Each day at feeding time, the mother mink would fashion five small patties from the scoop of ground meat given to her, and she would then call her offspring to eat. She never made four or six patties, but always five.

ILLUSTRATION FOR 237°

PROBABILITY AND STATISTICS

238° *A raffle.*   He bought tickets for every raffle of the lodge, but never won a thing. The lodge officers felt that they'd frame the next raffle so he'd win, for morale's sake. At the next selling of tickets he bought a number three. They decided to put nothing but number threes in a hat, blindfold him, have him stir up the tickets, and then pick one. He couldn't help but pick number three. Blindfolded he put his hand in the hat, stirred the tickets vigorously and picked one out—and out came $7\frac{1}{8}$.

—HARRY HERSHFIELD

239° *Theory versus reality.* If a coin falls heads repeatedly one hundred times, then the statistically ignorant would claim that the "law of averages" must almost compel it to fall tails next time. Any statistician would point out the independence of each trial, and the uncertainty of the next outcome. But any fool can see that the coin must be double-headed.—LUDWIK DRAZEK

240° *Theory versus reality again.* If there is a 50-50 chance that something can go wrong, then 9 times out of 10 it will.

—*Paul Harvey News,* Fall 1979.

241° *Conversion.* There's a 40 percent chance of snow tomorrow. That's 28 percent Celsius.

242° *A law of probability.* Chance favors only the prepared mind.—LOUIS PASTEUR

243° *Advantage of a near-zero probability.* There is a story told about a smart gas station owner who displayed a sign reading: "Your tank filled free if you guess how much it takes." He always had a long line of customers eager to take advantage of the offer, and of course the customer usually lost. When asked how his plan was working out, he said that about two years ago a fellow guessed right, but that it had cost him only $1.30. He said he was no longer getting customers asking for just a dollar's worth: everybody wants to guess on a fill-up.

244° *A classification.* There are three kinds of lies: lies, damned lies, and statistics.—BENJAMIN DISRAELI

In Item 308° of *Mathematical Circles Adieu,* this classification was incorrectly credited to Mark Twain.

245° *An interesting quote.*   Statistical thinking will one day be as necessary for efficient citizenship as the ability to read and write.—H. G. WELLS

246° *Advice.*   You should use statistics as a drunk uses a lamp post—for support rather than illumination.

—ANDREW LANG

247° *Characterizing a statistician.*   A statistician is a man who draws a mathematically precise line from an unwarranted assumption to a foregone conclusion.

248° *Statisticians placed end to end.*   If all statisticians were placed end to end, they would undoubtedly reach a confusion.

249° *Boys and girls.*   Statistics tell us that as far as growth is concerned, up to age twelve boys are about one year behind girls, during the ages twelve to seventeen, the boys gradually catch up, and from seventeen on it's neck and neck.

250° *Chances.*   An eager and energetic young man undergoing an interview for a position for which he had applied asked the manager of the firm what his chances would be of starting at the bottom and working his way to the top. "Very small," was the reply. "You see, we are in the business of digging wells."

251° *A statistic.*   It has been said that a person will exert himself 176 times as much to put something in an empty stomach as in an empty head.

252° *Economists versus statisticians.*   An economist is a man who begins by knowing a very little about a great deal and gradually gets to know less and less about more and more until he finally gets to know practically nothing about practically everything.

A statistician, on the other hand, begins by knowing a very great deal about very little and gradually gets to know more and more about less and less until he finally knows practically everything about nothing.

253° *Selling roof insurance.*   A salesman was trying to sell roof insurance to a home owner. "A hundred mile an hour gale," he said, "would rip every shingle off your roof." "No doubt it would," replied the home owner, "but we do not have such winds here. The highest wind ever recorded here was only ten miles an hour." "So," returned the salesman, "though you wouldn't lose all your roof shingles, you would still lose ten percent of them. Isn't that worth the insurance?"

254° *Statistics and a bikini bathing suit.*   Statistics are like a bikini bathing suit. What they reveal is suggestive, but what they conceal is vital.

255° *Further statistics.*   According to statisticians the average person spends at least one-fifth of his or her life talking. Ordinarily, in a single day enough words are used roughly to fill a 50-page book. In one year's time the average person's words would fill 132 books, each consisting of 400 pages.

ILLUSTRATION FOR 255°

256° *More statistics.*   After several practice drills, the pupils in a new school plant invited the superintendent and president of the school board to watch them in their fire drill. When the alarm rang, the three hundred pupils evacuated the building in one and one-half minutes.

The pupils went back to classes, proud and pleased. A while later when the noon whistle blew, the principal, still in possession of his stopwatch, made a test from idle curiosity. This time the building was cleared in less than one minute!

257° *There are figures and there are figures.*   At an orientation meeting in a small business college, the president of the college introduced to the student audience the director of admissions, a very attractive and shapely young woman, who presented some interesting enrollment statistics. When, upon finishing her part, the young lady was walking back to her seat on the platform, the president innocently exclaimed to the students, "I don't know about you, but figures like that really excite me."

258° *Of course.*   It is said that the average family has two and one-half children. This accounts for the large number of half-wits in the world.

259° *Statistics.*   I have five children. A friend, who had dabbled in the subject of statistics, remarked that if I should ever have a sixth child, it would have slanted eyes. Wondering if there was some genetical law I was unaware of, I asked why. "Because," replied my friend, "statistics has shown that every sixth child that is born is Chinese."

260° *A statistical figure.*   We are told that in 1960 there were 31 million children in the public schools of America. How many is that? If all those children were to march from the Atlantic Ocean to the Pacific Ocean and back again in rows of four, each row an arm's length from the one preceding it, the first row of children would have made the entire trip and returned to the Atlantic before the last row of children would have started the nationwide trek.

261° *Several heads better than one.*   The statistics instructor drew a line on the blackboard and turned to his class. "I'm going to ask each of you to estimate the length of that line." Rapidly he polled the class. Estimates ranged from 53 inches to 84 inches. The instructor put them down. Then he totaled them and divided the result by the number of students in the class. The average estimate, he announced, was $61\frac{1}{8}$ inches, although no one had given that exact figure.

Then he measured the line. It was $61\frac{1}{4}$ inches long.

This shows that several heads are better than one.

262° *A naive statistician.* A statistician administered a mathematics test to all six thousand people of a certain village and at the same time recorded the length of each person's feet. He was surprised to find a strong correlation between the mathematical ability and the foot size of the people. [Explain.]

FLAWED PROBLEMS

263° *"Wrong" answers trip testers.* In 1985 the Educational Testing Service gave students the opportunity to scrutinize (after the test) the questions, their answers, and the test key of recent PSAT and SAT tests. On each test, a student successfully challenged the examiners' choice of answer to a mathematical question. In acknowledging the errors, the ETS raised approximately 250,000 PSAT and 19,000 SAT scores. The embarrassing disclosures can be seen as a ray of mathematical sunshine—here is assurance that there are bright students who read problems carefully and are not willing to accept intended, but not stated, assumptions nor to accept answers dictated by authority alone. Here are the two questions challenged, together with the ETS answers and the challengers' answers.

Question: In the pyramids *ABCD* and *EFGHI* shown in the accompanying figure, all faces except base *FGHI* are equilateral triangles of equal size. If face *ABC* were placed on face *EFG* so that the vertices of the triangles coincide, how many exposed faces would the resulting solid have?

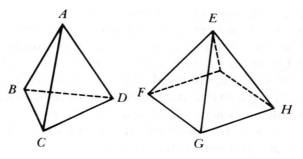

ILLUSTRATION FOR 263°

(A) Five   (B) Six   (C) Seven

(D) Eight   (E) Nine

ETS answer: (C).   *Challenger's answer:* (A)

Question: Which row in the list below contains both the square of an integer and the cube of a *different* integer?

Row A	7	2	5	4	6
Row B	3	8	6	9	7
Row C	5	4	3	8	2
Row D	9	5	7	3	6
Row E	5	6	3	7	4

ETS answer: Row B.   *Challenger's answer:* Row B and Row C.

264° *The challenger of PSAT Question 44.*   It was Daniel Lowen, a seventeen-year-old Florida student from Cocoa Beach High School, whose "wrong" answer to the geometry question in the PSAT test described above in Item 263° tripped the testers.

Daniel said he was struck by the apparent simplicity of Question 44 when he first read it. "It looked like a counting question, but it was the only one that didn't involve a formula," he said. "I figured that it had to be a trick question." He took another look and decided that when the two pyramids were joined four of the seven triangular faces merge into two quadrilateral faces of the new solid. "So I marked *Five* as the answer," he said. When he returned home that evening he made a model of the pyramids and confirmed his answer in his own mind.

When Daniel later received his scores, he learned that he had scored 48 out of 50 and that one of the two questions marked wrong was Question 44. The correct answer was given as *Seven.*

"It never entered my mind that they had made a mistake," Daniel said. "I figured that my model must have been inaccurate." He and his father, Douglas J. Lowen, a mechanical engineer who works for Rockwell International on the space shuttle, sat down to work out the problem mathematically.

"My dad tried to prove that I was wrong but he couldn't," Daniel said. "Then he came up with two different mathematical proofs that I was right."

The father then telephoned Educational Testing Service and followed up with a letter.

265° *Another flawed SAT problem.* There was another SAT test that contained a flawed problem. The question pictured a large circle adjacent to a small circle and stated that the radius of the large circle was three times that of the smaller one. It asked how many times the small circle would rotate while rolling once around the large circle.

Though the correct answer is 4, the testmakers thought the answer to be 3, and 4 did not even appear among the five choices.

266° *An unforgettable score.* While on the subject of test scores, it might be fitting to mention the following curious item, even though no flawed problems are involved.

Stephen Curran, a junior at Beloit College who received honorable mention in the Putnam Mathematical Competition held on Dec. 6, 1980, is not likely to forget his score or his rank in the competition. He placed forty-first in the forty-first annual Putnam competition with a score of 41.

—Noted in *Mathematics Magazine,* May 1981.

267° *Resurrection of an old problem.* Back in November of 1905, the *Bangor Daily News* ran the following problem: "A gentleman wishes to dig a trench, 100 feet in length, paying for the work $100—or an average of $1 a foot. He employs two laborers, agreeing to give the first laborer the sum of 75 cents per foot and the second $1.25 per foot. Now, how many feet—that is, what percent of the trench—must each laborer dig to earn $50?" The poser was signed "Mystified."

Seventy-two years later, in the December 2, 1977 issue of the *Bangor Daily News,* a Mrs. Annie Lawless of Clifton, Maine, re-opened discussion of the above problem by wondering if the problem, which baffled the experts of 1905, could perhaps be easily solved with the "new math" taught to school children today.

Mrs. Lawless apparently failed to realize that one respondent back in 1905 really saw through the problem, when he remarked: "Perhaps Mystified can answer this. If a cow costing $95 gives 20 quarts of milk a day, how high can a grasshopper jump without getting out of breath?"

Mrs. Lawless's optimism of the modern generation may be too great. On December 1, 1977 (the day before Mrs. Lawless reopened discussion of the old problem), the *Bangor Daily News* carried a story about the scores made by Florida youngsters in the literacy test that they must pass in order to get a high school diploma. Forty percent of them could not do simple arithmetic. One of those "simple" problems was to figure out how many gallon cans of paint it would take to cover a wall 12 feet high and 16 feet long if a gallon would cover 10 square yards.

268° *A recurring flawed problem.* There is a flawed problem that recurs frequently in newspapers and magazines, which runs as follows: Three men entered a hotel and rented a room for $30, or $10 apiece. Later, the manager of the hotel discovered that the men had inadvertently been overcharged, and that the cost of the room was only $25. Accordingly a page boy was given $5 and was instructed to return it to the three men. The boy, however, not being honest, pocketed $2 of the rebate and returned $3 to the men, who now paid $9 apiece for the room. Now the $27 paid by the men, and the $2 pocketed by the boy, add to only $29. What became of the extra dollar?

269° *The clock problem.* IQ tests sometimes contain flawed problems. Common among them is the clock problem: A clock reads twenty minutes to four. What time would the clock read if the hands of the clock are interchanged? The expected answer is: twenty minutes after eight.

The tester has failed to realize that time on a clock is really determined by the hour hand alone, and that the minute hand is a mere convenience. Thus, at twenty minutes to four, the hour hand is not on the 4, but is two-thirds of the way from the 3 to the 4. Interchanging the two hands would place the hour hand on the 8; the minute hand would then have to be on the 12 and could not be two-thirds of the way from the 3 to the 4.

The clock problem does suggest an interesting related one: What times give correct clock readings when the hands are interchanged?

ILLUSTRATION FOR 269°

270° *The induction problem.*   On IQ tests one often finds questions like: "What is the next term in the sequence 1, 4, 9, 16 . . . ?" The expected answer is "25," the fifth term in the sequence whose nth term is n^2. Actually, the fifth term might be· any number whatever, say π. Thus, the sequence whose nth term is

$$f(n) = n^2 + (n - 1)(n - 2)(n - 3)(n - 4)(\pi - 25)/24$$

has 1, 4, 9, 16, π for its first five terms.

QUADRANT FOUR

From an optical illusion
to a remarkable factorization

RECREATION CORNER

271° *An optical illusion.* The number on the right below looks larger, but the one on the left can be proven to be twice as big.

ILLUSTRATION FOR 271°

—MICHAEL A. STUEBEN

272° *Another optical illusion.* As you look at the square below, the black dot in the center seems to disappear.

ILLUSTRATION FOR 272°

—MICHAEL A. STUEBEN

273° *A mnemonic for pi.* A pretty mnemonic for recalling the decimal expansion of π to ten places is the following one in Spanish:

Sol y Luna y Mundo proclaman al Eterne Autor del Cosmos!
3. 1   4   1   5   9   2   6   5   3   6

The last digit, 6, is rounded up from the actual value 5, since the next three digits are 8, 9, 7.

274° *A mnemonic for* e.   Here is a mnemonic for recalling the first few digits in the decimal expansion of *e*:

He studied a treatise on calculus
2.   7   1   8   2   8

275° *The triangle mnemonic.*   Mnemonic devices have been found useful in helping beginning pupils solve certain types of problems. For example, if we have three quantities *a*, *b*, and *c* such that $a = bc$, and we want to solve for some one of them in terms of the other two, we place the three quantities in a triangle as indicated in the accompanying figure. Then we place a finger over the quantity we are seeking and perform the indicated operation on the remaining two quantities.

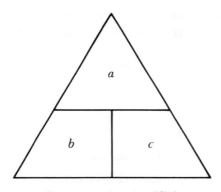

ILLUSTRATION FOR 275°

Examples of such relations are:

distance = rate × time
volts = amperes × ohms
watts = volts × amperes
mass = density × volume
force = mass × acceleration

276° *Geometric and algebraic viewpoints.*   Can you move one match to make a perfect square?

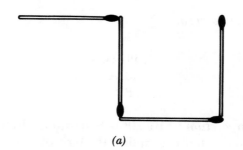

(a)

Answers:

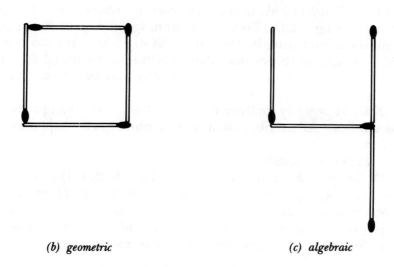

(b) geometric   *(c) algebraic*

ILLUSTRATION FOR 276°

277° *Iff.* John Horton Convoy, the British game expert, extended the familiar "iff" (if and only if) to "onnce" (once and only once), "onne" (one and only one), "whenn" (when and only when), and so on. Richard K. Guy has told of an amusing converse that appeared in Canada at the 1981 *Banff Symposium on Ordered Sets.* Someone sported a T-shirt with

BANF AND ONLY BANF

lettered on it.

278° *3° below zero.*

$$0$$
Ph.D.
M.A.
B.S.

279° *A prediction.*   In 1966 Martin Gardner's mystical Dr. Matrix predicted* that the millionth digit of π would be 5, since in the King James Bible, third book, chapter 14, verse 16, the magical number 7 appears and the seventh word has 5 letters. In 1974 J. Guilloud and his associates in Paris calculated π to 1,000,000 decimals, using a CDC 7600, in a running time of 23 hours and 18 minutes. Surprisingly, the millionth digit of π turned out to be 5. (The millionth decimal place, excluding the initial 3, is 1.)

—Pointed out by GEORGE MIEL.

280° *A proof by mathematical induction.*   We shall prove the theorem $P(n)$: All numbers in a set of n numbers are equal to one another.

I. $P(1)$ is obviously true.

II. Suppose k is a natural number for which $P(k)$ is true. Let $a_1, a_2, \ldots, a_k, a_{k+1}$ be any set of $k + 1$ numbers. Then, by the supposition, $a_1 = a_2 = \ldots = a_k$ and $a_2 = a_3 = \ldots = a_k = a_{k+1}$. Therefore $a_1 = a_2 = \ldots = a_k = a_{k+1}$, and $P(k + 1)$ is true.

It follows that $P(n)$ is true for all natural numbers n.

281° *Another proof by mathematical induction.*   We shall prove the theorem

$P(n)$: If a and b are any two natural numbers such that $\max(a,b) = n$, then $a = b$.

I. $P(1)$ is obviously true.

II. Suppose k is a natural number for which $P(k)$ is true. Let a and b be any two natural numbers such that $\max(a,b) = k + 1,$.

---

*In Martin Gardner, *New Mathematical Diversions from Scientific American*. New York: Simon and Schuster, 1966.

and consider $\alpha = a - 1$, $\beta = b - 1$. Then $\max(\alpha,\beta) = k$, whence, by the supposition, $\alpha = \beta$. Therefore $a = b$ and $P(k + 1)$ is true. It follows that $P(n)$ is true for all natural numbers n..

282° *Miscellaneous methods for proving theorems.* *Reductio ad nauseum,* proof by handwaving (sometimes called the Italian method), proof by intimidation, proof by referral to nonexistent authorities, the method of least astonishment, the method of deferral until later in the course, proof by reduction to a sequence of unrelated lemmas (sometimes called the method of convergent irrelevancies), and finally, that old standby, proof by assignment.
                    —*Rome Press 1979 Mathematical Calendar.*

283° *The square root of passion.* Steve Shagan, in his book *City of Angels* (New York: Putnam's Sons, 1975, p. 16) says, "But you can't make arithmetic out of passion. Passion has no square root." To this Alan Wayne replied, "On the contrary, one can show that the alphametric

$$\sqrt{\text{PASSION}} = \text{KISS}$$

has a unique solution."

284° *Observed on an automobile bumper sticker.*

Mathematicians are number $-e^{i\pi}$.

285° *A brief list of mathematical phrases with their most common meanings.*

1. Clearly: It can be shown in half a day.
2. Details omitted: The author couldn't establish the result.
3. An idiot: Anyone less mathematically clever than we.
4. It is not difficult: It is very difficult.
5. It is easily seen that: It is false.
6. Ingenious proof: A reference the author could understand. Often called trivial.
7. Interesting: Dull.

8. A genius: Anyone smarter than we.
9. Obviously: It can be shown in three pages.
10. Similarly: We can show it but are too lazy.
11. Without loss of generality: We proved only the easiest case.
12. Well-known (result): A result whose reference cannot be located.

—MICHAEL A. STUEBEN

286° *An octagon.*   Some second graders were identifying geometric forms held up by their teacher. When she showed them a square, they shouted "Square." A triangle was just as easy. And almost all knew what a rectangle was. Then she held up an eight-sided shape.

"What is this one?"

To a child, they told her, "A stop sign."

287° *The new mathematics.*   Standard mathematics has recently been rendered obsolete by the discovery that for years we have been writing the numeral five backward. This has led to reevaluation of counting as a method of getting from one to ten. Students are taught advanced concepts of Boolean algebra, and formerly unsolvable equations are dealt with by threats of reprisals.—WOODY ALLEN

288° *Take mathematics.*   How can you shorten the subject? That stern struggle with the multiplication table, for many people not yet ended in victory, how can you make it less? Square root, as obdurate as a hardwood stump in a pasture—nothing but years of effort can extract it. You can't hurry the process.

Or pass from arithmetic to algebra; you can't shoulder your way past quadratic equations or ripple through the binomial theorem. Instead, the other way; your feet are impeded in the tangled growth, your pace slackens, you sink and fall somewhere near the binomial theorem with the calculus in sight on the horizon. So died, for each of us, still bravely fighting, our mathematical training; except for a set of people called "mathematicians"—born so, like crooks.—STEPHEN LEACOCK

289° *The revelation.*   I had a feeling once about Mathematics—that I saw it all. Depth beyond depth was revealed to me—the Byss and Abyss. I saw—as one might see the transit of Venus or even the Lord Mayor's Show—a quantity passing through infinity and changing its sign from plus to minus. I saw exactly why it happened and why the tergiversation was inevitable—but it was after dinner and I let it go.—WINSTON CHURCHILL

290° *Hoggatt's solution of a numbers game.*   There was a popular numbers game played back in the 1940s that engaged many mathematicians across the country. It had originated in a problem in *The American Mathematical Monthly*. The game was to express each of the numbers from 1 through 100 in terms of precisely four 9's, along with accepted mathematical symbols of operation.

Some years later this game evolved into what seemed a much more difficult one; namely, to express the numbers 1 through 100 by arithmetic expressions that involve each of the ten digits 0, 1, . . . , 9 once and only once. This game was completely and brilliantly solved when Verner Hoggatt, Jr. discovered that for any nonnegative integer n,

$$\log_{(0 + 1 + 2 + 3 + 4)/5}\{\log_{\sqrt{\sqrt{\ldots\sqrt{(-6 + 7 + 8)}}}}9\} = n,$$

where there are n square roots in the second logarithmic base. Notice that the ten digits appear in their natural order, and that, by prefixing a minus sign if desired, Hoggatt had shown that *any integer*, positive, zero, or negative, can be represented in the required fashion.

Another entertaining numbers game of the period was that of expressing as many of the successive positive integers as possible in terms of not more than three π's, along with accepted symbols of operation.

291° *A unit of measure for beauty.*   Since mathematics is now being applied to esthetics, perhaps the following definition of a unit of measure for beauty is in order: A *millihelen* is that precise amount of beauty just sufficient to launch a single ship.

ILLUSTRATION FOR 291°

292° *Daffynitions.*

1. Deduce: The lowest card in the deck.
2. Minimum: A very small mother.

HAVE YOU HEARD?

RECREATIONISTS in the area of mathematics occasionally indulge in the sport of manufacturing absurd stories about fictitious mathematical discoveries. The following items are representative of this genre.

293° *Set theory.* Georg Cantor's mother was discussing with her son the arrangements for an upcoming dinner party. "What we need," said Mrs. Cantor, "is a new set of dinnerware to be used just for times of entertainment." She proceeded to elaborate that with each dinner plate there should be a salad plate, a cup and saucer, and so on. Her son, possessed of an abstract mind and intrigued by the above-mentioned one-to-one correspondences, later repaired to his study and began to work out the basic principles of set theory.

294° *Extra-set theory.* Mitch Coleman, a Chicago-based free-lance writer, who, it has been claimed, "just doesn't add up," has researched the origin of a concept known as *extra-set theory*. Coleman says that this extension of ordinary set theory was first proposed in 1933 by a little-known American mathematician named Carl Weinberg, who devised it to keep track of his car keys, which he was always misplacing. As a practical application of the theory, Weinberg kept an extra set of car keys taped to the inside of his gas cap.

295° *The Möbius strip.* A number of versions of the true discovery of the Möbius strip have been offered. One version maintains that the mathematician August Ferdinand Möbius once took a vacation at the seashore. He found himself so pestered at night with flies that he secured a strip of paper sticky on both sides. Giving the strip a half-turn and pasting the two ends together, he hung the resulting loop from a rafter in the bedroom of his vacation cottage. His improvised flycatcher worked well and he slept undisturbed by flies. Awaking one morning after a fine night's rest, his eye fell upon the flycatcher hanging over his bed and he noticed, to his surprise, that the strip had only one side and only one edge. Thus was born the famous Möbius strip.

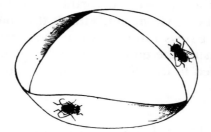

ILLUSTRATION FOR 295°

296° *The energy equation.* Several accounts have been given of the flash that led Albert Einstein to the famous energy equation $E = mc^2$. According to Gary Larson (the clever creator of "The Far Side"), the great scientist was at his small blackboard writing out and rejecting one form of the energy equation after another. He tried $E = mc^3$, $E = mc^7$, $E = mc^4$, $E = mc^{10}$, and so on. As he was successively crossing out these forms, his cleaning lady dashed in. Ignoring the presence of Einstein, she snapped her dust cloth and feather duster about and quickly straightened objects on the desk. Stepping back to survey her work, she commented aloud that things looked better now that she had *squared* them away. Hearing the word *squared*, a gleam swept over Einstein's face and, with satisfaction at last, he wrote on the blackboard the formula $E = mc^2$.

297° *The alternate field theory.*   It seems that field theory originated a lot earlier than previously supposed. Credit appears to go to a Chinese farmer Clung Long, who flourished about 1160 in one of the southern provinces of China. He discovered that· planting a farm field on alternate years and letting it lie fallow on the other years led, in the long run, to better crops. To distinguish this early field theory from the later algebraic field theory, it has been deemed wise to call it the *alternate field theory.*

—A. A. AARON

298° *A complete ordered field.*   It wasn't until close to a century after the discovery of the alternate field theory (see Item 297°) that field theory received further refinement. In 1243, a Chinese farmer Long Clung introduced the concept of planting vegetables in neat rows rather than in the former helter-skelter fashion and of using the entire field rather than just spots of it. These concepts led to the theory of complete ordered fields.

—A. A. AARON

299° *Unified field theory.*   Albert Einstein never succeeded in finding a unified field theory. This problem, however, was actually solved years earlier by socialist farming practices. By the socialist system, individual farmers combined their separate wheat fields into a single large cooperative field. By this device, the failure of a crop could be blamed on one another.

—RUTH STEARNS

300° *The origin of topology.*   It is sometimes said that topology came into existence when someone failed to see any difference between a square and a circle. Recent explorations in a cave near Barbeston, Spain have revealed a wall painting wherein it seems that this failing occurred as early as the first millenium B.C. The drawing, herewith reproduced, shows a square horse looking round. This drawing has been heralded by historians as one of the great archeological finds of recent times.

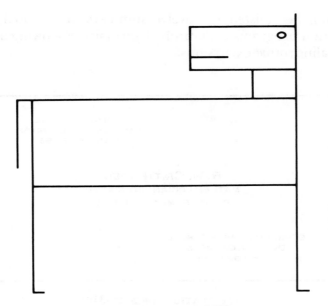

ILLUSTRATION FOR 300°

—B. T. FISCHER

301° *The theory of logs simplified.* A farmer, Jeremiah Stone of Vermont, back in 1894, sold 18-inch logs for woodstoves. He had two piles of logs, one labeled *natural logs* and the other *common logs*. He sold the natural logs for twice the cost of the common logs. Complaints from his clientele to the rural council led to the council looking into the matter. The council could discover no essential difference between the two types of logs and ordered the farmer to adopt a single price for both kinds, thus leading to an appreciated simplification in the theory of logs.

—T. B. HENDERSON

MR. PALINDROME

R. W. CRITTENDEN, a longtime and outstanding teacher of high school mathematics, is a palindrome enthusiast. He is also interested in teaching problem solving via the methods of George Pólya, whom he assisted for twelve summers in National Science Foundation programs at Stanford University. Following are a few of his palindromic excursions.

---

578 Gellert Drive
San Francisco, CA 94132
(415) 66-161-66

R. W. Crittenden
MATHEMATICIAN - EDUCATOR
PALINDROME SPECIALIST

SERRAMONTE HIGH SCHOOL
Daly City, CA 94015
(415) 992-9555

Illustration for 302°–310°

302° *An interesting date.*   On the eighth of February of this year (1982) I was reminded of the palindromic nature of the date 2/8/82. Investigating further, I noted that 2882 is the forty-fourth pentagonal number. I also discovered that the forty-fourth heptagonal number is palindromic, namely 4774. Likewise with the forty-fourth nonagonal number, 6666, and the forty-fourth 11-gonal number, 8558. Alas, the forty-fourth 13-gonal number is not palindromic.—R. W. Crittenden

303° *A large palindromic square.* $637832238736 = (798644)^2$.—R. W. Crittenden

304° *A triangular palindrome.* $15051 = (173)(174)/2$.
—R. W. Crittenden

When this number appeared on the odometer of his car, Crittenden stopped the car and took a photo of the car's dashboard.

305° *An interesting palindrome.* The palindromic number 919 is interesting because

$$9^3 + 1^3 + 9^3 = 1459 \quad \text{and} \quad 1^3 + 4^3 + 5^3 + 9^3 = 919.$$

—R. W. Crittenden

306° *A personalized license plate.* Seeing the IXOHOXI of Item 351° in *Mathematical Circles Adieu,* I went to the California Department of Motor Vehicles and requested a personalized plate: IXOHOI. I am enclosing a photo of the plate.

ILLUSTRATION FOR 306°

—R. W. Crittenden

307° *Prime palindromic postage from Crittenden.*

R. W. CRITTENDEN
578 Gellert Drive
San Francisco, CA 94132

Professor Howard Eves
Box 251, RFD 2
Lubec, Maine   04652

(a)

(b)

ILLUSTRATION FOR 307°

308° *Prime palindromic representations of 20-cent post-age.* I have found 22 *prime* palindromic representations of 20-cent postage.—R. W. CRITTENDEN

With the occasional change of the first-class postage rate, such as the 1985 change from 20 cents to 22 cents, Crittenden has the task of revising his list of prime palindromic representations.

ILLUSTRATION FOR 308°

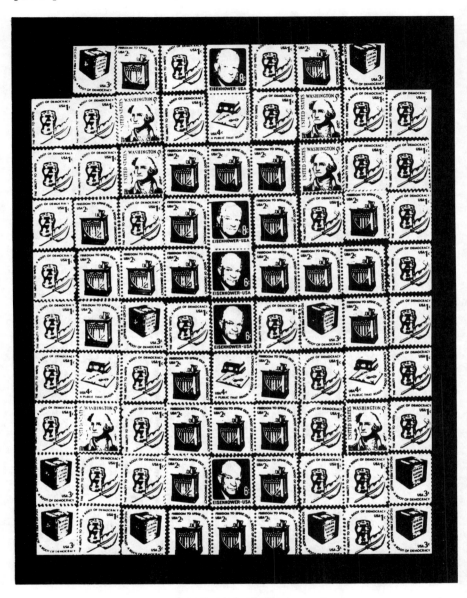

ILLUSTRATION FOR 308° *(continued)*

ILLUSTRATION FOR 308° *(continued)*

309° *United States cities with prime palindromic zip codes.* A few United States cities with *prime* palindromic zip codes are:

75557	Boston, Texas
31013	Clinchfield, Georgia
12421	Denver, New York
35753	Hytop, Alabama
93239	Kettleman City, California
96769	Makaweli, Hawaii
74047	Mounds, Oklahoma
15451	Lake Lynn, Pennsylvania
97879	Troy, Oregon

The following also have prime palindromic zip codes:

Amawalk, New York	Great Valley, New York
Bloomingburg, New York	Markleton, Pennsylvania
Comstock, New York	West Sunbury, Pennsylvania
Eagle Bay, New York	Tylersburg, Pennsylvania
Peakville, New York	Madera, Pennsylvania

Adairsville, Georgia	Elkins, Arizona
Auburn, Georgia	Kemp, Oklahoma
Avera, Georgia	Iowa Park, Texas
Baxley, Georgia	Mexia, Texas
Bascom, Florida	Westfield, Texas
Crane Hill, Alabama	Stafford, Texas
Graham, Alabama	Placedo, Texas
Perdido Beach, Alabama	Nevada City, California

Sorry, I can't find any in Maine!—R. W. CRITTENDEN

310° *For Crittenden's files.*   In the November 1980 issue of *Crux Mathematicorum* appeared a table of all 93 five-digit and all 668 seven-digit palindromic primes. The calculation was done on a PDP-11/45 at the University of Waterloo. The calculation time was slightly more than one minute. A particularly attractive nine-digit palindromic prime is the number 345676543, given by Léo Sauvé, editor of the above journal, who states that there are 5172 nine-digit palindromic primes.

EXAMPLES OF RECREATIONAL MATHEMATICS BY THE MASTER

CHARLES W. Trigg, esteemed and admired by the entire mathematical community, has long been king of the mathematical recreationists. His clever wit and agile mind have furnished recreational material for a large number of journals for many years and have left readers agape with wonder at his astonishing inventiveness. In this section we offer a mere sample, and largely one-sided at that, of Trigg's skill; one could easily devote an entire booklet to mathematical recreations devised by him.

One must not get the idea that Trigg has done little else in mathematics than construct clever recreational items. That is far from the case. He has authored, in solo and sometimes with coauthors, a large number of serious mathematical papers. His name has appeared in essentially all of the American journals of mathematics and in many foreign ones. He is one of the country's leading and most prolific problemists and for a time served as the editor of the problem section of *The Mathematics Magazine*.

Now a professor emeritus and dean emeritus of Los Angeles City College, he is fondly remembered there for his skillful and inspiring teaching and his great administrative talent. He has written a charming book, *Mathematical Quickies* (McGraw-Hill, 1967, reprinted by Dover, 1985), that should be in every mathematics teacher's library. An unbelievably busy and productive person has been Charles W. Trigg.

311° *The game of aggregates.*   Familiar to practically everyone are expressions like:

a covey of quail	a herd of cattle
a gam of whales	a pride of lions
a school of fish	a gaggle of geese
a bouquet of pheasants	an exhilaration of larks

Nouns of multitude to describe certain aggregations seem originally to have been applied, as in the above examples, to groups of animals; and some of these applications are almost as old as the English language itself.

Several years ago it became a game to apply nouns of multitude to aggregates other than groups of animals. James Lipton in 1986 gave an extensive wider list, including, for example:

an impatience of wives	an obeisance of servants
a flush of plumbers	a shush of librarians
a rash of dermatologists	a galaxy of astronomers
a column of accountants	a fifth of Scots

To these, later gamesters have added, among many others:

a pile of nuclear physicists	a press of journalists
a grid of electrical engineers	a set of pure mathematicians

a litter of geneticists	a magazine of editors
a stack of librarians	a wing of ornithologists
a peck of kisses	a complex of psychologists

In all these lists the area of mathematics was essentially neglected. To correct this grave omission, in 1982 Charles W. Trigg produced a list of forty well-chosen examples, among them being:

a mapping of cartographers	a printout of computers
a calculation of arithmeticians	a rationalization of fractions
a congruence of number theorists	an incomprehensibility of symbols
a correlation of statisticians	a quibble of logicians
a circle of geometers	a foundation of axioms
an aberration of angle trisectors	a restriction of conditions
a distortion of topologists	a paucity of even primes
an integration of analysts	a plethora of digits

Since Trigg's initial efforts, the mathematical list has grown considerably.

312° *A contribution from the third grade.* The third grade teacher was carefully explaining the different words in the English language used to describe a group. Thus a group of sheep is a *flock* and a group of quail is a *covey*. Then she asked for the names of groups of other animals. When she came to camels, a child timidly suggested, "A carton?"

313° *The game of mama-thematics.* Another recreational "sport" that spread through the mathematical community a few years ago is that known as *mama-thematics*. The following examples, all given by master player Charles W. Trigg, clarify the nature of the game:

Mother Einstein to Albert: "You see, I always told you to spend less time playing that violin and to think more about relatives."

Mother Khayyam to son Omar: "Your algebra and astronomy may have been helped by that jug of wine, but they didn't bring in much bread."

Mother to Archimedes: "I do declare, you must be in your second childhood, playing in that sandbox all day!"

Mama-san to son Y. Yoshino: "In this computer age, I'm glad that you are loyal to your abacus. You can count on it."

Napier's mother to her son: "What's all this talk I hear about your bones and logs? Anybody'd think you had a wooden leg."

Mrs. Newton to son Isaac: "Don't come running to me for sympathy. If I've told you once, I've told you a hundred times not to sit under that apple tree."

Mama Ceva to son Giovanni: "Lucky you did not set out to be a fisherman, the way you are always getting your lines crossed."

Mama to Maria Agnesi: "If you don't stop walking around in your sleep, people will think that *you* are the witch!"

Mrs. Babbage to son: "Charles, you're just too lazy to do arithmetic in your head, so you try to build a machine to do it for you. What kind of an example is that for my grandchildren?"

Mother to Euclid: "I should be proud of your writing, but it's all Greek to me."

Mother to Leonardo Fibonacci: "So that's why you wanted those Easter bunnies."

Mother to Cardan: "Why can't you get along with that Tartaglia lad?"

Madame de Fermat to Pierre: "If the margin is too small, why not use the flyleaf?"

Mrs. Shanks to son William: "You should have used a computer."

Madame de Buffon to her son: "What has suddenly made you so clumsy, dropping needles all over the place?"

Mother to Lobachevsky: "If parallel lines diverge, why haven't there been more wrecks on the Trans-Siberian railroad?"

Mother to Archimedes: "Mother didn't know it was lost, but now that you have found it why all the screaming?"

Mrs. Dedekind to son Richard: "And I thought you wanted that knife to whittle with."

Mrs. Cantor to son Georg: "I'm very glad that you gave up singing for mathematics."

Mrs. Conway, about son John: "Ever since he was a boy he's been a game kid."

Madame Pascal, about son Blaise: "When he got involved with those gamblers and into that triangle, I was afraid Blaise would become blasé."

—CHARLES W. TRIGG

119

314° *A holiday permutacrostic.*   Each of the following phrases is a permutation of letters of a mathematical term. The first letters of the terms spell the name of a holiday.

(1)	A TOURING LATIN	(9)	VIOLET CAR
(2)	HE GOT NAP	(10)	A LIMIT IN FINES
(3)	CAN RENT GAT	(11)	WIN ON A TEN
(4)	PAIN IN EAR	(12)	RED GIANT
(5)	TIME IN SACK	(13)	DON SUIT IN CITY
(6)	TEN BUDS	(14)	THAT GIRL IS MONA
(7)	TREE TAGS	(15)	I DARE TONY
(8)	'TIS GREEN		

Solution:

(1)	Triangulation	(9)	Vectorial
(2)	Heptagon	(10)	Infinitesimal
(3)	Arctangent	(11)	Newtonian
(4)	Napierian	(12)	Gradient
(5)	Kinematics		
(6)	Subtend	(13)	Discontinuity
(7)	Greatest	(14)	Antilogarithms
(8)	Integers	(15)	Y-ordinate

—CHARLES W. TRIGG
School Science and Mathematics, Dec. 1966.

315° *A holiday message in a permutacrostic.*   Each of the following phrases is a permutation of the letters of a mathematical term. The first letters of these terms in order spell out our holiday message.

(1)	A ST. LOUIS MENU	(7)	SIT CATS, SIT
(2)	LET ME YEARN	(8)	DATE GUARD
(3)	APPEAR TO MIX	(9)	TORO
(4)	G-MEN SET	(10)	ALL ARE QUIET
(5)	NO TIRADE	(11)	SAXON PINE
(6)	RUN LAME	(12)	TORE HEM

(13) ON AIR TRAIL	(15) STAG TREE
(14) NOT A TONI	(16) SAT RIGHT

Solution:

(1) Simultaneous	(8) Graduated
(2) Elementary	(9) Root
(3) Approximate	(10) Equilateral
(4) Segment	(11) Expansion
(5) Ordinate	(12) Theorem
(6) Numeral	(13) Irrational
(7) Statistics	(14) Notation
	(15) Greatest
	(16) Straight

—CHARLES W. TRIGG
School Science and Mathematics, Dec. 1962.

316° *A coffee expert.*   Before continuing with more of Trigg's contributions to recreational mathematics, we pause to insert five interesting items concerning Trigg himself.

In 1916 Trigg became an Industrial Fellow at the Mellon Institute of Industrial Research, University of Pittsburgh, directed to develop a process for manufacturing soluble coffee. The group remodeled a Detroit brewery building to produce instant coffee under the trade names of Minute Coffee and Coffee Pep. The process was efficient, the product was excellent, but inadequate financing led to bankruptcy. Trigg moved to Los Angeles in 1924.

During the 1916 to 1924 interval, Trigg published 32 articles on the chemistry of coffee, tea, and spices in the *Tea and Coffee Trade Journal,* together with 132 short notes, and 32 editorials. One of the articles, "Health and Happiness in Spices," was the first prize article in the 1923 American Spice Trade Association National Contest and was widely reprinted internationally. Trigg also wrote the signed chapters, "The Chemistry of the Coffee Bean" and "Pharmacology of the Coffee Drink" in W. H. Ukers' definitive 1922 book, *All About Coffee.*

ILLUSTRATION FOR 316°

317° *An unusual bid to fame.*   On February 5, 1955, sub-stituting for the Dean of Admissions, Trigg registered motion picture actress Katherine Grant Crosby into classes at Los Angeles City College before a battery of television and still cameras. Columbia Pictures chose this time to announce the charming lady's first pregnancy. As a result, LACC and Charles W. Trigg got more publicity than either received from a single event before or since. The staff of the *L. A. C. C. Collegian* assembled the 449 clippings that they had received from *Life, Time,* and newspapers in 28 states and New Zealand into a string book, which they presented to Trigg.

318° *Great problemist.*   Trigg's first mathematical publication was the solution of Problem 1207 in the April 1932 issue of *School Science and Mathematics.* Since that time, 3577 solutions submitted by Trigg to challenge problems in various journals have been acknowledged as correct; 1011 of these have been published. In addition, 630 of his problem proposals and 53 quickies have appeared in various problem sections. Furthermore, over the years, 312 of his mathematical articles and notes, 197 book reviews, 21 letters to the editor, and 36 items akin to MathMADics have been distributed among 31 mathematical periodicals, domestic and foreign.

319° *A tribute.*   When Trigg retired he received the following tribute from Professor Nathan Altshiller Court: "Your renown as a problemist is not based on the volume of your production alone. You endow your contributions with a quality which is rare, namely, wit. Yes, mathematics may be witty, and you provide the proof."

320° *Paper folder.* A fascinating branch of recreational mathematics concerns itself with paper folding, and here Charles W. Trigg has produced some gems—from folding a regular hexagon into a regular tetrahedron to a host of intriguing folding problems involving an ordinary correspondence envelope.

It was shortly after Trigg's discharge from the Navy and return to Los Angeles City College in 1945 that he became interested in the geometry of paper folding. This led to the construction of polyhedra with cardboard panels and rubber bands. Some of the more colorful hung as mobiles in his office, and others were periodically exhibited in display cases in the Administration Building. Just prior to his retirement, the Engineering Department presented him with a colorful diploma awarding him the degree of P.F.P.D. (Polyhedra Doctor in Paper Folding).

321° *Names of mathematicians in timely anagrams.*

(1) A CALM RUIN
(2) AND RAG
(3) MA, JUAN RAN
(4) ARM ONCE
(5) I'M THERE

(6) I POLL NO U.S.A.
(7) PAR ON ICE
(8) NO HARPS
(9) AIM HIS CLUB
(10) A HILL TOP

Solution:

(1) Maclaurin
(2) Argand
(3) Ramanujan
(4) Cremona
(5) Hermite

(6) Apollonius
(7) Poincaré
(8) Raphson
(9) Iamblichus
(10) L'Hôpital

—CHARLES W. TRIGG
Mathematics Magazine, Mar.-Apr. 1961.

322° *Initialed numbers.*   A quiz devised by Will Shortz, editor of *Games* magazine, involves taking a phrase containing a number, reducing some of the words to their initials, and challenging the reader to reconstruct the phrase. For example: "26 = L. of the A." came from "26 letters of the alphabet," and the origin of "12 = E. on an O." is "12 edges on an octahedron." In the spirit of Shortz, you are asked to complete the following mathematical phrases.

(1)	6 = F. in a F.	(6)	6 = S. P. N.
(2)	2 = O. E. P.	(7)	7 = T. O. P.
(3)	3 = M. O. P.	(8)	8 = S. E. C.
(4)	4 = L. E. S.	(9)	11 = O. T. D. P. P.
(5)	5 = S. on a P.	(10)	12 = M. A. N.

Answers: (1) 6 feet in a fathom; (2) 2, only even prime; (3) 3, minimum odd prime; (4) 4, least even square; (5) 5 sides on a pentagon; (6) 6, smallest perfect number; (7) 7, third odd prime; (8) 8, smallest even cube; (9) 11, only two-digit palindromic prime; (10) 12, minimum abundant number.

—Charles W. Trigg
Journal of Recreational Mathematics, Vol. 15(3), 1982–83.
© 1982, Baywood Publishing Co., Inc.

323° *Matching doubles.*   Can you match each of the following expressions in the left-hand column with one of the digit pairs on the right?

Banquet	00
Cry of an impatient golfer	11
Hospital ward	22
Naturals	33
Emphatic denial	44
Short and fat	55
Bet on 7 races and lost on 6	66
Successive hat tricks	77
The house number	88
Abbreviated ballet skirt	99

Answers:

Banquet—eighty ate	88
Cry of an impatient golfer—Fore! Fore!	44
Hospital ward—sixty sicks	66
Naturals—on the first throw in dice play, seven or eleven	77
Emphatic denial—Nein! Nein!	99
Short and fat–five by five	55
Bet on 7 races and lost on 6—won one	11
Successive hat tricks—three goals scored by one player in each of two games	33
The house number—the double zero in roulette	00
Abbreviated ballet skirt—tutu	22

—CHARLES W. TRIGG

Journal of Recreational Mathematics, Vol. 13(4), 1980–81.
© 1980, Baywood Publishing Co., Inc.

324° *Are you letter-perfect?* You can find out if you are letter-perfect by putting in each blank space the letter of the English alphabet that is described or defined by the word, number, or phrase that follows the blank space. Find the resulting message.

_____ slope

_____ base of natural logs

_____ distance from the in-center to a side of a triangle

_____ quotient of two successive terms of a G.P.

_____ ordinate

_____ constant

_____ two hundred

_____ $abc/4\Delta$

_____ $\sqrt{-1}$

_____ $(a + b + c)/2$

_____ the degree of an equation

_____ five hundred

_____ side of a triangle opposite $\sphericalangle\alpha$

_____ orthocenter

_____ 1, 7, 13, 19 . . .
_____

_____ genus of a surface (topology)

_____ axis

_____ representation of an integer

_____ eccentricity of a conic

_____ looks like a primitive· cube root of unity

_____ one hundred and sixty	_____ 150
_____ noon	_____ spherical excess
_____ high grade	_____ coefficient of the
_____ not N, E, W	squared term in the
	general quadratic
_____ commonly the first	_____ one-half the space di-
term of a progression	agonal of a cube

Answer:

merry CHRisTMAS anD a HA.P.pY New YEaR

[Authority for Y = 150, T = 160, and H = 200 is *The Random House Dictionary* (Unabridged).]

—CHARLES W. TRIGG
Journal of Recreational Mathematics, Vol. 15(2), 1982–83.
© 1982, Baywood Publishing Co., Inc.

325° *A tale with the homonymic powers of two.*

Arriving at the golf club long before their scheduled starting time, the two female foursomes went to the dining room and the

2^3 eight ate.

Later, on the course, eager to drive off, the trailing

2^2 four called "Fore!" for the fore four to move on.

At the end of the regulation game, there was a tie for first place so they awarded the prize

2^1 tutu to two, too.

However, the contestants insisted upon playing until

2^0 one won.

—CHARLES W. TRIGG
Journal of Recreational Mathematics, Vol. 14(1), 1981–82.
© 1981, Baywood Publishing Co., Inc.

326° *Dig the ten digits.*

1 is the digit of ego,
2 of the blissful pair,

3 of marriage gone on the rocks, and
4 is the sign of the square.
   5 cards is a hand in a poker game.

6 they say is perfection,
7 come 11, invocative rhyme,
8 the ball you get behind, and
9 often is curfew time.
   0 is actor Mostel's first name.

—CHARLES W. TRIGG
Mathematics Teacher, Apr. 1969.

327° *An invariant determinant.*

E 1016 [1952, 328]. *Proposed by Norman Anning, University of Michigan*

Find the element of likeness in: (a) simplifying a fraction, (b) powdering the nose, (c) building new steps on the church, (d) keeping emeritus professors on campus, (e) putting B, C, D in the determinant

$$\begin{vmatrix} 1 & a & a^2 & a^3 \\ a^3 & 1 & a & a^2 \\ B & a^3 & 1 & a \\ C & D & a^3 & 1 \end{vmatrix}$$

Solution by C. W. Trigg, Los Angeles City College. The value, $(1 - a^4)^3$, of the determinant is independent of the values of B, C, and D. Hence, operation (e) does not change the value of the determinant but merely changes its appearance. Thus the element of likeness in (a), (b), (c), (d), and (e) is only that the appearance of the principal entity is changed. The same element appears also in: (f) changing the name-label of a rose, (g) changing a decimal integer to the scale 12, (h) gilding the lily, (i) whitewashing a politician, and (j) granting an honorary degree.

—*American Mathematical Monthly*, Feb. 1953.

328° *Some integrations.* Along the lines of the well-known integration

$$\int d(\text{cabin})/\text{cabin} = \log \text{cabin} + C = \text{houseboat}$$

we have:

(1) $3\int (\text{ice})^2 d(\text{ice}) = \text{ice cube} + C = \text{iceberg}$

(2) $2a\int \text{real } d(\text{real}) = \text{a real square} + C = \text{movie hero}$

(3) $t\int du = c + ut = \text{wound}$

(4) $\int d(\text{art}) = \text{art} + C = \text{Cart}$

(5) $\int d(\text{wall}) = \text{wall} + C = \text{dike}$

(6) $p\int d(\text{lane}) = p(\text{lane}) + C = \text{hydroplane}$

(7) $4\int dt = 4t + C = \text{sea fort} = \text{battleship}$

(8) $\int_1^2 dx/x = \ln 2 = \text{lean-to}$

(9) $10\int_0^t dx = 10t = \text{tent}$

(10) $8\int_0^t dx = t(4)(2) = \text{tea for two}$

(11) $a\int_0^1 dw/\sqrt{1 - w^2} = a\pi/2 = \text{half a pie}$

(12) $\int_0^{\text{board}} \cos y \, dy = \text{sin board}$

(13) $\int_0^{\text{ema}} \cos x \, dx = \text{sin ema} = \text{movie}$

(14) $\quad -\displaystyle\int_{\pi/2}^{salad} \sin x \, dx = \cos$ salad $=$ romaine salad

(15) $\quad \displaystyle\int_{0}^{hide} \sec^2 z \, dz = \tan$ hide

(16) $\quad -\displaystyle\int_{\pi/2}^{springs} \csc^2 v \, dv = \cot$ springs

(17) $\quad \displaystyle\int_{-1}^{champagne} \sec u \tan u \, du = \sec$ champagne $+ \sec 1$

$= $ dry champagne $+$ dry one

(18) $\quad \displaystyle\int_{up}^{down} \sinh x \, dx = \cosh$ down $- \cosh$ up $= 2 \cosh$ down

$=$ to quiet down

(19) $\quad 8\displaystyle\int_{i\text{-tooth}}^{eye} x \, dx = $ eye(4)eye $+$ tooth(4)tooth

(20) $\quad 4\displaystyle\int_{-m}^{m} dx = (m8 + 8m)/2 = $ mate ate'm, too

—CHARLES W. TRIGG
Mathematics Teacher, Dec. 1970.

329° *The famous names game.* The famous names game became popular a dozen or more years ago, and Charles W. Trigg promptly extended it to the names of mathematicians. We here give a sample of Trigg's work; a more complete list can be found in the *Journal of Recreational Mathematics*, Summer, 1974.

Pity SALMON, the poor POISSON!
Where has J. D. GERGONNE?
Despite his cutting ways, was DEDEKIND?
Who dares to pull ROBERT'S. BEARD?
Did his method of exhaustion sour him so that he was ANTIPHON?
As active currency in the U.S., do you ever expect to

C. GOLDBACH?

Would his preferred avocation B. TAYLOR?

If he found a FIBONACCI bonanza, would VERNER HOGGATT?

If his girl got lost, would J. A. H. HUNTER? Or would TODHUNTER?

In perplexity, when he ran his fingers through his hair did the WHITEHEAD-RUSSELL?

He drew a BEDE on the BALL, but in making the POOL shot, which cushion did LEON BANKOFF?

How well can RONALD C. READ? Does DALE SEYMOUR? Which is the RUDERMAN?

Which provides the best means of driving through the maze of mathematics, a FORD or an OLDS?

Does A. C. BRADBURY his mistakes, or does he carry D. CROSS?

For JOHN PECKHAM or BACON is necessary to maintain his GROSS, rotund BOMBELLI, even though the DERNHAM is HIGH and RICH.

Which has the richer overtones, ABEL of Norway or thE. TEMPLE BELL?

Which R. BOYD watchers most likely to C. WREN, EAGLE, CRANE, PEACOCK, HERON, or OWLES? Beware the FOWLER who is an ARCHER with SPEARS and TRAPP!

Was COMTE DE BUFFON a probability hunt for a needle in a pi?

If we do not appreciate their theorem, will STEINER-LEHMUS? Is DERRICK LEHMER?

It is hard to tell about RAPHSON, whether he CANTOR won't put on the EUCLID.

—CHARLES W. TRIGG

Journal of Recreational Mathematics, Vol. 7(3), 1974.

© 1974, Baywood Publishing Co., Inc.

MISCELLANEA

330° *Ronald Reagan and mathematics.*

NEW YORK (AP)—President Reagan in a letter he wrote 30 years ago confessed that as a student, "I was very poor at mathematics and took only what was absolutely required."

The letter will be auctioned April 15 at Swann Galleries, Inc., in Manhattan, one of only a few Reagan letters to reach the market.

The letter is dated Jan. 4, 1952, when Reagan was still a movie actor and before he entered politics. It was addressed to Tom Tweddale, of Fort Worth, Texas, a high school student who had asked advice on how to become a sports announcer—one of Reagan's earliest jobs.

While admitting he had a weakness in math, Reagan advised the youth, "And again I say—get a college education."

Swann estimates the letter will bring between $3,000 and $4,000.

—*Bangor Daily News,* Mar. 9, 1982.

331° *Re the metric system.*　The rest of the world is probably right to bully us into adopting the metric system since it's convenient for everybody to tell the same lies. But let us not give it the benediction of the scientific community.

—KENNETH E. BOULDING

332° *President Garfield and the metric system.*　In 1879, at a meeting in the Old South Church of Boston, a group of zealots organized The International Institute for Preserving and Perfecting Weights and Measures. The aim of the institute was to strive for a revised system of units of measure to conform to sacred standards believed inherent in the construction of the Great Pyramid of Egypt and ceaselessly to combat the "atheistic metrical system" of France. United States President James A. Garfield became an enthusiastic supporter of the institute, but when offered its presidency, he declined the honor.

333° *It all depends.*　When James A. Garfield was president of Hiram College, a student inquired about the mathematics course he had to take. "Can't the course be shortened? I could never learn all this." Garfield replied, "It depends upon what you want. When God wants an oak, it takes him a hundred years; but when he wants a pumpkin, it takes only three months."

ILLUSTRATION FOR 333°

334° *The good of it all.*   A mathematics professor started his first class by presenting a strong case for the course, whereupon a student asked, "Can you prove to me that this course is all that good?"

The professor reached into his desk, took out his lunch, and from it extracted a banana. Peeling the banana he ate it as the class watched. On finishing, he turned to the student and asked, "Do you know what that particular banana tasted like?"

"No," came the reply, "Only the one who ate it can tell that."

"So it is with this mathematics course," concluded the professor. "You must taste it yourself."

335° *There's always an easier way.*   When asked what it was like to set about proving something, the mathematician likened proving a theorem to seeing the peak of a mountain and trying to climb to the top. One establishes a base camp and begins scaling the mountain's sheer face, encountering obstacles at every turn, often retracing one's steps and struggling every foot of the journey. Finally, when the top is reached, one stands examining the peak, taking in the view of the surrounding countryside—and then noting the automobile road up the other side!

—ROBERT J. KLEINHENZ

336° *A perilous act.*   A Harvard mathematics student who might cite, much less use, a proof from a book by the great Yale analyst Pierpont, did so at his peril. In all likelihood the proof would not be accepted.

337° *The author Schnitt.*   As a high school teacher of some seven years, it occurred to me that even though I had recommended Bell's *Men of Mathematics* to many students, I had never read it cover to cover. So I began to do so. As students would enter my office, I would often share with them the latest things I had learned from this remarkable book. One day Don, a senior and perhaps the best-read high school student of mathematics I have ever known, came into the office. I began showing off the newest things I had gained from the book and proceeded to give him a trivia test. I asked him to recall the author of the following quotation:

> The heart of Dedekind's theory of irrational
> numbers is his concept of the "cut"
> . . . *(Schnitt)*

He frowned and said he didn't know.

"Ha!" I retorted. "Have you read the book or not?" (Indeed, he had read it several times.) Then I informed him that the author of the quotation was Schnitt.

He looked puzzled and asked to see the page. Then he looked at me with an expression of frustration and said softly, "Mr. Leonard, *Schnitt* is the German word for cut!"

—WILLIAM LEONARD
Two-Year College Mathematics Journal, Jan. 1979.

338° *Boiling water.*   A mathematics student approached his professor and made the following request: "You are given a stove and a pan filled with water on an adjacent table. Your task is to boil the water. Explain how you would solve the problem." The professor thought for a moment, and then proceeded, "First look in the pan; if the water is boiling, we are done. Otherwise. . . ."

–THOMAS R. DAVIS
Two-Year College Mathematics Journal, Jan. 1979.

339° *Obvious.*   One of my professors at UCLA was going to prove a theorem of the form, "*A* if and only if *B*." He started writing before the bell rang at the beginning of the class, wrote (and droned) steadily, filling the chalkboards on three walls with small writing, continued past the dismissal bell and stopped when the next batch of students began invading the room. At that point he had proved only "If *A*, then *B*." He surveyed the three boards full of cramped notes and announced breezily that "in the other direction it's equally obvious!"

—DAVID E. LOGOTHETTI
Two-Year College Mathematics Journal, Mar. 1979.

340° *Again, obvious.*   As a young Ph.D. at Harvard with one of those appointments long on title and short on salary, Burt Rodin decided to listen to a lecture by one of the older and more famous professors. Part way through his presentation, the great man stopped talking and just stood there. After several minutes of nearly painful silence, a student less diffident than the others raised his hand and said, "Er, ah, Herr Doktor Professor, Sir— what are you doing?" His reply was, "I was just trying to decide whether what I was going to say next is obvious or not; I've decided it is, so I won't say it."

—DAVID E. LOGOTHETTI
Two-Year College Mathematics Journal, Mar. 1979.

341° *Mathematicians at work.*   D. R. Curtiss proposed that mathematics might be publicized at the Chicago World's Fair. Automobile trailers were just coming into fashion. He suggested that a trailer equipped with two desks and chairs be parked on the grounds of the fair with a large sign asserting, "Mathematicians at Work."

342° *A faultless rejection.*   The following letter is offered as a consolation to the many authors who have submitted manuscripts for publication, only to receive a letter with the phrase "we regret. . . ."

After a British writer submitted for publication a paper on the economy to a Chinese journal, he received the following rejection letter (quoted in the *World Business Weekly*):

"We have read your manuscript with boundless delight. If we were to publish your paper it would be impossible for us to publish any work of a lower standard. And as it is unthinkable that, in the next thousand years, we shall see its equal, we are, to our regret, compelled to return your divine composition, and to beg you a thousand times to overlook our short sight and timidity."

—*Mathematics Magazine,* May, 1981.

343° *The applicability of mathematics.*   There is nothing mysterious, as some have tried to maintain, about the *applicability* of mathematics. What we get by abstraction from something can be returned!

—R. L. Wilder
Introduction to the Foundations of Mathematics.

344° *The power of memory.*   We all know that books burn— yet we have the greater knowledge that books cannot be killed by fire. People die, but books never die. No man and no force can abolish memory.—Franklin Delano Roosevelt

345° *A perfect notation.*   It is India that gave us the ingenious method of expressing all numbers by means of ten symbols, each symbol receiving a value of position as well as an absolute value; a profound and important idea which appears so simple to us now that we ignore its true merit. But its very simplicity and the great ease which it has lent to computations put our arithmetic in the first rank of useful inventions; and we shall appreciate the grandeur of the achievement the more when we remember that it escaped the genius of Archimedes and Apollonius, two of the greatest men produced by antiquity.—Pierre-Simon Laplace

346° *An indictment.*   Mathematical communication in journals has become so concise and devoid of any intuitive background that it has been claimed that many published mathematical papers have been read by only three people—the author, the editor, and a referee.

347° *A nasty taste in the mouth.*   Once, on a plane ride, a pleasant and loquacious gentleman took the seat next to me and started up a conversation. Things went swimmingly until he asked me, "What do you do for a living?" "I teach mathematics," I replied. He gave a sudden convulsive cough, scrambled to his feet, murmured "Pardon me," and bolted to the back of the plane with his hand over his mouth. When he emerged, he quietly dropped into a seat several rows behind me.

348° *An editorial comment.*   In the manuscript for the fifth edition of my *An Introduction to the History of Mathematics,* I had

written, in connection with the poorer Newtonian fluxional notation and the better Leibnizian differential notation, "The English mathematicians, though, clung long to the notation of their
leader." The copy editor deleted the word "long," and wrote in
the margin, "I thought Clung Long was a Chinese mathematician,
not English."

349° *The St. Augustine quote.*   When I was a student a theo-
logical friend amused himself by quoting at me St. Augustine's
alleged injunction to beware of mathematicians lest they lead one
to damnation. I have seen this quoted again quite recently. Mathematics has a poor press, but this particular derogatory statement
is a canard: when St. Augustine wrote "mathematician" he meant,
like many classical authors, "astrologer." The actual text reads,
*"Bono christiano sive mathematici sive quilibet inpie divinantium, maxime
dicentes vera, cavendi sunt, ne consortio daemonicorum animam pacto
quodam societatis inretiant."* This may be rendered more or less as
follows: A good Christian must beware of astrologers as well as
of those soothsayers who make predictions by unholy methods,
and most especially when their predictions come true; he must
guard against their having arranged to ensnare his soul by de-
ceiving him through association with demons.

(*De genesi ad litteram,* Book II, Chapter xvii; in *Corpus Scrip-
torum Ecclesiasticorum Latinorum,* Vol. 28, Part I (Vol. 3, Part 1 of
Augustine's works), edited by J. Zycha, Prague-Vienna-Leipzig,
1894, pp. 61–62.)

—RALPH P. BOAS
American Mathematical Monthly, Feb. 1979.

[For the usual St. Augustine quote see Item 149° in *Mathematical Circles Adieu.* Mathematicians should be pleased for the
light Professor Boas has thrown upon the proper understanding
of the quote. Actually, St. Augustine had a high regard for mathematics and once remarked that the two most perfect things he
could think of are ethics and mathematics.]

350° *Student resourcefulness.*   One hears complaints about
students' lack of resourcefulness. Back around 1940, one of my

calculus students complained that he couldn't do an assigned problem because he didn't know what a horizon was.

I had a large class one year in a course for nonmathematics majors, most of whom didn't like it very much. On the course evaluation forms, one of them, in answer to the question "What most interested you about this course?," wrote "The professor's collection of bow ties."

—RALPH P. BOAS
Two-Year College Mathematics Journal, Jan. 1979.

351° *Mathematics compared to an oyster.* When an irritation is set up inside an oyster by a foreign particle, the oyster quietly solves the problem by exuding a substance that allays the friction, covering the sore spot, and miracle of miracles, the substance hardens, forming a pearl. Oysters that have never been irritated, never had a problem. No problem, no pearl. Similarly in mathematics, the solution of irritating problems often leads to the creation of mathematical gems.

ILLUSTRATION FOR 351°

352° *A philosophy of learning and teaching.*

Free Mathematics
(Available here Mon. thru Fri.)

But you must bring your own container, and *you* must fill it with much or little according to its capacity and the amount of work that you are willing to do. The *learning assistant* (sometimes euphemistically called a "teacher") will provide expertise, advice, guidance, and will set an example. But in the final analysis it is *you* who must do the work needed for *your* learning—as, indeed, I must do for mine. Theories to the contrary, however well intentioned, are mistaken.

As a poet once said:

> Let us then be *up* in doing,
>   With a heart for *any* "fate,"
> Still achieving, still pursuing—
>   Learn to *labor* and to *wait*.

and again:

> It matters not how straight the gait,
>   How charged with punishments the scroll;
> *I* am the captain of my fate,
>   *I* am the shepherd of my soul!

Here it is—this wonderful stuff called *mathematics*. If you want it, come and get it. If you don't want it, kindly step out of the way—so as not to impede the progress of those who do. The *Choice* is yours. May God bless you, keep you safe, and reward you according to your deservedness. Those who chose *for* mathematics, please step this way.—L. M. CHRISTOPHE, JR.

353° *The checker king.*   The present (1985) world checker champion has remained unbeaten for thirty years. Despite a $5,000 challenge from a checker organization, no computer program has been devised that will beat him. There is no other game that has had a comparable champion.

Who is this remarkable reigning king of the checkerboard? He is Marion Tinsley, a fifty-eight-year-old professor of mathematics at Florida A & M University in Tallahassee. Tinsley graduated from Ohio State University, where, in 1957, he secured a Ph.D. in mathematics, with a special interest in combinatorial analysis.

The Encyclopedia of Checkers declares, "Marion Tinsley is to checkers what Leonardo da Vinci was to science, what Michelangelo was to art, and what Beethoven was to music." He has a photographic memory and the ability to look thirty moves ahead and accurately picture the positions on the board.

For Tinsley, checkers and mathematics complement one another. He sees many parallels between the two endeavors. Elegantly solving a checker problem, he says, is like deftly demonstrating a mathematical proof.

354° *Cistern problems.* Many type-problems of elementary algebra have enjoyed long histories. In Item 13° of *In Mathematical Circles,* we described a type-problem, still familiar today, which has been traced back in time at least to the Rhind papyrus of about 1650 B.C. Another type-problem with a long history is the so-called cistern problem, originally concerned with filling cisterns by means of pipes having given rates of flow.

The cistern problem seems to have first appeared, in definite form, in Heron's *Metrica* of about A.D. 100. It is next found in the works of Diophantus of about A.D. 275 and among the Greek epigrams attributed to Metrodorus of about A.D. 500. Soon after, it became common property in both the East and the West. It was found in the list of problems attributed to Alcuin (*ca.* 800), in the Indian classic *Lilāvati* of Bhāskara (*ca.* 1150), and in subsequent Arabian arithmetics. When books began to be printed, the cistern problem was among the stock problems of such early writers as Tonstall (1522), Frisius (1540), and Recorde (*ca.* 1540).

Originally, the cistern problems reflected an observation of daily life; anyone living along the Mediterranean coast saw cisterns that were filled by pipes of various diameters. But there is an interesting law of textbook writers—that it is quite all right to steal from one another with almost no scruples provided the theft is thinly veiled. Accordingly, the cistern problem went through a number of metamorphoses.

Thus, starting in the fifteenth century, we find variations involving a lion, a dog, and a wolf, or other animals, eating the carcass of a sheep. In the sixteenth century we find further variations involving men building a wall or a house—problems of the form: "If A can do a piece of work in 4 days, and B in 3 days, how long will it take if both men work together?"

In a work of Frisius (1540), the problem becomes a ridiculous drinking problem: "A man can drink a cask of wine in 20 days, but if his wife drinks with him it will take only 14 days—how long would it take the wife alone?" Under the growth of commerce we also find the case of a ship with three sails, by the aid of the largest of which a voyage can be made in 2 weeks, with the next in size in 3 weeks, and with the smallest in 4 weeks—find the time if all three sails are used. Here the problem, that in its original cistern form had a practical aspect, has become unrealistic, as it ignores the matter of one sail blanketing another and the fact that the speed of the ship is not proportional to the area of sail. Probably the height of absurdity was reached when one writer proposed: "If one priest can pray a soul out of purgatory in 5 hours, while it takes a second priest 8 hours, how long will it take if the two priests pray together?"

Since the solution of a cistern problem involves a special procedure, it is quite certain that problems of this genre will continue to be found among the *story problems* of our elementary algebra textbooks.

355° *Von Neumann and a trick problem.* There is an old "trick" problem that is resurrected every now and then. The problem concerns two fictitious trains and a bee. The trains, which travel at given constant rates, simultaneously leave New York and Chicago, traveling toward each other on a single straight track. At the moment the train from New York sets out, a bee, flying at a given rate exceeding the rate of each train, flies toward the Chicago train. When the bee meets the Chicago train, it reverses its direction and flies toward the New York train. When it meets the New York train, it reverses its direction and flies toward the Chicago train. The bee keeps up its to-and-fro flight between the two trains until it is squashed when the trains finally meet head on. Knowing the distance between New York and Chicago, how far in all has the bee flown?

The problem is a "trick" problem because an unthinking solver will often try to find the distance of the bee's flight by computing the length of its successive flights between the two trains, and then summing the resulting convergent series. But there is a much easier way to solve the problem. First find how long the trains travel before meeting. This time, multiplied by the bee's rate of flight, will give the total distance traveled by the bee.

It is said that the trains-and-bee problem was once proposed to the mathematical genius John von Neumann, who almost immediately gave the correct answer. "Ah," remarked the poser of the problem, "you know the simple way of solving the problem." And he went on to explain to von Neumann the complex method that many unthinking solvers follow. "Oh," exclaimed von Neumann, "is there a simpler method?"

Quite likely the above incident never occurred, but had it occurred, would von Neumann's brilliantly almost instantaneous solution of the problem by the long and complex method outweigh his stupidity in failing to see the short and easy method?

John von Neumann was born in Budapest in 1903 and was soon recognized as a scientific prodigy. He took his doctorate in Budapest in 1926, migrated to America in 1930, and in 1933 became a permanent member of the Institute for Advanced Study at Princeton. He already had an international reputation for his contributions to logic, the foundations of mathematics, operator theory, quantum theory, and game theory. He did much to determine the direction of a great deal of twentieth-century mathematics. His work was remarkably bold and original, and he had an almost uncanny ability to foresee many coming important areas of research. During World War II he engaged in scientific and administrative work related to the hydrogen and atomic bombs and to long-range weather forecasting. He died of cancer in 1957. It may well be that time will register him as the most brilliant genius of the present century.

ILLUSTRATION FOR 355°

356° *Farkas Bolyai's many interests.*   Farkas Bolyai, the father of János Bolyai, was a man of many interests. In addition to his work in mathematics, he spent considerable time composing tragedies. In middle age he translated Pope's *Essay on Man* into Hungarian. Besides his devotion to poetry, he loved music and played the violin. A peculiar hobby of his was the construction of ovens of unusual designs. It is said that the domestic economy of Transylvania was revolutionized by one of his ovens possessing a special arrangement of flues. About his room were discarded oven models, interspersed with favorite violins. Here and there hung portraits—one of his friend Gauss, another of Shakespeare, whom he called the "child of nature," and a third, of Schiller, whom he called the "grandchild of Shakespeare." It has been reported that he frequently compared the earth to a muddy pool, wherein the fettered soul waded until death came, and a releasing angel set the captive free to visit happier realms.

357° *Adolphe Quetelet and statistics.*   The Belgian astronomer and mathematician Adolphe Quetelet (1796–1874) was a pioneer in the field of statistics and was the first to make a statistical breakdown of a national census. In 1829 he analyzed the first Belgian census, noting the influence of age, sex, season, occupation, and economic status on mortality. The bearing of such an analysis on life insurance is obvious.

Quetelet created the concept of "the average man." His statistical studies convinced him that crime in a given population is, to a certain extent, mathematically predictable.

Among Quetelet's most ardent converts to the value of statistical studies was Florence Nightingale (1820–1910), who believed that "to understand God's thought, we must study statistics, for these are the measure of His purpose."

358° *Doubling an estimate.*   Stanislaw Ulam, in an address given several years ago, estimated that about 100,000 new theorems are published annually. A more careful later estimate, made by two younger mathematicians who were in Ulam's audience, doubled this estimate.

359° *A salacious book?*   In one of his undergraduate classes, Professor Brinkmann mentioned a book by Blaise Pascal, *Essai sur les passions que se font dans l'amour,* but did not recommend his students borrow the book from the library. "It's probably out," whispered H. Wexler, then a student in the class and later to become chief of the U.S. Weather Bureau.

360° *A remarkable factorization.*   The Minimite friar Marin Mersenne (1588–1648), a French number theorist who maintained a constant correspondence with the greatest mathematicians of his time, is today chiefly remembered in connection with the so-called *Mersenne primes*, or prime numbers of the form $2^p - 1$, p prime, which he discussed in a couple of places in his work *Cogitata physico-mathematica* of 1644.

It is common today to represent the number $2^p - 1$ by M_p. In his work of 1644, Mersenne stated, without proof, that M_{251} is composite. It was not until the nineteenth century that mathematicians finally proved Mersenne correct, by finding that M_{251} contains both 503 and 54,217 as prime factors. However, complete prime factorization of M_{251} was not achieved until February 1984, when two researchers employing a thirty-two-hour search on a CRAY-supercomputer found that

$$2^{251} - 1 = 503 \times 54,217 \times 178,230,287,214,063,289,511$$
$$\times\ 61,676,882,198,695,257,501,367$$
$$\times\ 12,070,396,178,249,893,039,969,681.$$

It is now known that M_p is prime for the following 29 exponents p:

2, 3, 5, 7, 13, 17, 19, 31, 61, 89, 107, 127, 521, 607, 1279, 2203, 2281, 3217, 4253, 4423, 9689, 9941, 11213, 19937, 21701, 23209, 44497, 86243 and 132049,

and for no other $p < 50,000$.

M_{132049}, which contains 39751 digits, is today the largest known prime number. It was found with the aid of a CRAY-supercomputer at the Lawrence Radiation Laboratory in California in 1983 by David Slowinski.

Since even perfect numbers (it is believed there are no odd ones) are of the form

$$2^{p-1}(2^p - 1),$$

where $2^p - 1$ is prime, it follows that the value $p = 132049$ yields a perfect number, the twenty-ninth and largest perfect number known today.

EPILOGUE

Ninety minutes of humor
Sixty minutes of teaching tips
Twenty-five minutes of logic

SOME MATHEMATICAL HUMOR, IN MINUTE DOSES

1′ A boy borrowed a book on algebra from the school library. Three days later he returned it with the plaintive protest, "This book tells me more about algebra than I want to know."

2′ Many who teach mathematics need the prayer of the old Scot, who feared decay from the chin up: "Lord, keep me alive while I'm still living."

3′ "Let me illustrate the value of this theory," said the absentminded mathematics professor as he erased the blackboard.

4′ Arithmetic is neither fish nor beast; therefore, it must be fowl.

5′ Two teenage girls were discussing their problems. One said, "You shouldn't be discouraged. Today, there is a man for every girl, and a girl for every man. How can you improve on such an arrangement?"

"I don't want to improve on it," retorted the other. "I just want to get in on it."

6′ The teacher, in the last week of school, was trying desperately to give her class an impression of fractions that would last through the summer. She told them they could think of fractions at home as well as in school and gave such examples as "half a sandwich," "a quarter of a pie," and "tenth part of a dollar." At that point one little boy "caught on" and proudly contributed, "My father came home last night with a fifth."

7′ Some mathematics students are like blotters. They soak it all up, but get it backwards.

8′ A gambler's seven-year-old son, when asked to count in school, promptly responded: "1, 2, 3, 4, 5, 6, 7, 8, 9, 10, jack, queen, king."

9' "Tommy, what is 'nothing'?" asked the teacher. " 'Nothing' is a balloon with its skin peeled off," Tommy replied.

10' The mathematics teacher loaded his class down with enough problems to keep them engaged for several hours. After fifteen minutes, when the teacher was settled comfortably in his swivel chair, his reverie was marred by, "Sir, do you have any more problems?"

Somewhat aghast, the teacher queried, "Do you mean you have finished all those I assigned?"

"No," answered the student, "I couldn't work any of these, so I thought I might have better luck with some others."

11' "Is your mathematics teacher very strict?"

"Is he? You remember Smitty? Well, he died in class, and the teacher propped him up until the lecture was over."

12' If all the students who fall asleep in college math classes were laid end to end, they would be more comfortable.

13' "In most mathematics departments," said the chairman, "half the committee does all the work, while the other half does nothing. I am pleased to announce that in our department it is just the reverse."

14' Confronted with a serious problem, Sandy, a college freshman, sent her mother a special delivery airmail letter reading:

"Dear Mother: Please send me $40 for a new dress immediately. I've had six dates with Tommy and have worn each of the dresses I brought to college. Have another date next Saturday night and must have another dress for the occasion."

Her mother solved the problem in a reply via Western Union: "Get another boy friend and start over."

15' A math professor sent his son to a rival college, and the lad came home at the end of the first year jubilantly announcing that he stood second in his math class.

"Second?" said the father. "Second? Why weren't you first?"

Filled with determination, the boy plowed into his math books and returned home from his sophomore year with top honors in mathematics. His father looked at him silently for a few minutes, then shrugged his shoulders and grumped, "At the head of the class, eh? Well, that college can't have much of a math department!"

16' A convention has been characterized as the confusion of the loudest talking delegate multiplied by the number of delegates present.

17' A convention is a succession of 2's. It consists of 2 days, which are 2 short, and afterward, you are 2 tired 2 return 2 work and 2 broke not 2.

18' One student confessed that the only thing he learned in his math class was how to sleep sitting up.

19' Nobody ever got hurt on the corners of a square deal.

20' Being completely baffled by a particular question in the mathematics midterm exam, a college student finally inserted "This rings no bell" below the question.

When the papers were returned, the student found that the professor had written a note of his own. It read: "Ding-Dong—page 83."

21' A student was asked by his history of mathematics professor to name the principal contribution of the Phoenicians. The answer? "Blinds."

22' This note appeared on a high school math exam, "Views expressed in this paper are my own and not necessarily those of the textbook."

23' An ill-prepared college student taking a math exam just before Christmas vacation wrote on his paper, "Only God knows the answers to these questions. Merry Christmas!"

The professor graded the papers and wrote this note: "God gets 100, you get 0. Happy New Year!"

24' Mathematics teachers are doing quite well financially these days, I hear; I have just learned of a mathematics teacher who started poor at the age of 20, and retired with a comfortable fortune of $50,000. This sum was accumulated through industry, economy, conscientious effort, perseverance, and the death of an uncle who left him $49,990.

25' One of the tragedies in mathematics is the murder of a beautiful theory by a brutal gang of facts.

26' "There's no sense in teaching the boy to count over 100," said Mr. Newrich to his son's tutor. "He can hire accountants to do his bookkeeping."

"Yes, sir," murmured the tutor, "but he'll want to play his own game of golf, won't he?"

27' Whenever two people meet, there are really six people present. There is each man as he sees himself, each man as the other person sees him, and each man as he really is.

28' "Willie," said the teacher, "if fuel oil is selling for $1 a gallon and you pay your distributor $200, how many gallons will he bring you?"

"About 190 gallons," answered Willie, after some thought.

"Why Willie, that isn't right," said the teacher.

"No, Ma'am, I know it ain't," said Willie, "but they all do it."

29' "You can't come in here and ask for a raise just like that," said the superintendent. "You must work yourself up."

"But I did," replied the young mathematics teacher, "look, I'm trembling all over."

30' "I'll have to have a raise, sir," said the mathematics teacher to the superintendent, "or I'll have to leave the profession. There are three companies after me."

"What three?" demanded the superintendent.
"Light, telephone, and gas," was the reply.

31' "Are you working hard on your trigonometry course?"
"Yes, I'm constantly on the verge of mental exertion."

32' When dessert was served, young Jimmy finally reached what threatened to be his limit of expansion. He reached for his belt buckle and explained, "Guess I'll have to move the decimal point two places."

33' There is the mathematics professor who is dieting—he wants to win the nobelly prize.

34' It seems that today's three R's are rockets, radar, and radioactive materials.

35' "Well, then," the father went on, "if you have one dollar and I have one dollar, and we exchange, we each have one dollar. But if I have an idea and you have an idea and we exchange, we each have two ideas. Right?"
His daughter is still trying to figure it out—mathematically.

36' *Math professor:* Have you heard about the new do-it-yourself idea?
Student: No. What is it?
Professor: It's called homework.

37' It was almost time for high school to let out in the spring and a boy was asking his Dad for an advance on his allowance. Dad asked for the reason and the son replied, "Well, our mathematics teacher is leaving our school and the class wants to give him a little momentum."

38' Epitaph: Here lies Napier's bones.

39' The pure mathematician noisily complained to his neighbor that the neighbor's children had made footprints in his new concrete sidewalk. "Don't you like children?" asked the neighbor. "Oh, yes, I like them well enough," said the pure mathematician, "but in the abstract, not in the concrete."

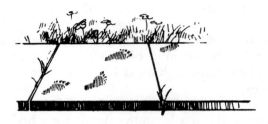

ILLUSTRATION FOR 39'

40'   Hitting a child across the knuckles with a ruler when he bungled his multiplication tables had one advantage over modern child psychology. It made the child smart.

41'   After the arithmetic prizes were passed out in the fourth grade, Rusty's mother asked him if he had received an award. "No," he replied, "but I got a horrible mention."

42'   Telephone conversation between two seventh grade boys: "All right, page 11, problem 8—what answer does your Dad get for that one?"
"He can't solve it so he's checking with a CPA."

43'   A safety sign read, "School—Don't Kill a Youngster." Beneath was scrawled, "Wait for the Math Teacher."

44'   A junior high math teacher jokingly told his pupils on report day that if their parents wouldn't let them come home because of bad grades, they could all come to his home to live. That evening when he went home for dinner, he found forty-one pupils sitting on his porch.

45'   Tommy, undergoing serious chastisement for his poor mark in mathematics, asked: "Well, Dad, what do *you* think is wrong with me—heredity or environment?"

46′   The young applicant hopefully presented himself to the interviewer of an engineering firm. "How were your math grades in college?" asked the interviewer. "They were all below water," replied the applicant cryptically. "Below water? What do you mean?" queried the interviewer. "They were all below C-level," admitted the applicant reluctantly.

47′   "Professor," said the curious student, "will you explain to me the theory of limits?"

"Well, young man, let us assume that you have called on an attractive young lady. You happen to be seated at opposite ends of the living room. You move toward her half the distance; then you move half the remaining distance toward her; again you decrease the distance between you by 50 percent. Continue this for some time. Do you get the idea? Theoretically, you will never reach the girl. On the other hand, you will soon get close enough to her for all practical purposes."

48′   "Do you know," said the mathematics instructor to a lazy student, "that your head is to your body as an attic is to a house—the highest point and the most empty." (An example of a proportion.)

49′   The president of the school board was a bit curious about a cutie just hired to assist with the teaching of arithmetic. "Can she multiply and divide, add and subtract?" he asked.

"No," said the superintendent, "but she certainly can distract."

50′   A mathematical-physicist was being driven to the auditorium in a metropolitan area by his chauffeur. The scientist, who was to deliver an address that evening, was rehearsing his speech as the limousine rolled along. The admiring chauffeur remarked, "I sure wish I could speak like you do and hold the attention of large audiences." "Well, it's not too difficult," said the scientist. "Why don't you take my manuscript, look it over, and give the address tonight—I'll drive; and by the way, no one knows me here. The audience will never know the difference." The chauffeur finally consented. Strangely enough, he got along very well but was taken aback when someone in the audience called for a question-and-answer period. When a technical question was addressed to him, he said, "Well, it is really very simple; just to prove how simple it is, I think I'll let my chauffeur answer it."

51′   A mathematician once arrived to speak in Baltimore and found only a smattering of listeners present. Not at all perturbed, he remarked, "Altogether there are only fourteen present; however, I am certain that I am privileged to speak to the fourteen most intelligent people in all Baltimore."

52′   A merchant became curious when week after week a local mathematics teacher came in and bought several brooms. Finally he sought an explanation.

"Well," said the teacher, "I'm selling them to my neighbors and friends for a dollar each."

"Look, man," protested the merchant, "you can't go on doing that. You're paying me $1.25 each for the brooms."

"That's right, I know," conceded the teacher, "but it beats teaching."

53′   The butcher noticed a pistol in a mathematics teacher's pocket when the teacher stopped in the butcher shop one day, and the butcher reported the observation. A policeman nabbed the teacher when his car stalled in traffic. "Come out with your hands up, and no funny business," ordered the policeman. Searching the teacher, the policeman found the gun—a water pistol. "I have a dozen like it at home," the mathematics teacher explained. "I've been confiscating them from my students."

54′   A nearsighted mathematics teacher was rapidly losing his temper. "You at the back of the class—what is the quadratic formula?"

"I don't know."

"Well, then, can you tell me how many roots a quadratic has?"

"I don't know."

"I taught that last Friday. What were you doing last night?"

"I was drinking beer with some friends."

The teacher gasped, and his face went almost purple. "You

have the audacity to stand there and tell me that! How do you expect to pass your examination?"

"Well, I don't. I'm an electrician, and I just came in here to fix the light."

55′ The day before last year's eclipse of the moon, the teacher announced to his sixth grade class that they should watch the total eclipse at 9 o'clock the following evening. He described it as one of the most wonderful shows that nature ever offers and stressed the fact that it would be free to everyone to enjoy. When he had finished, a world-weary eleven-year-old asked resignedly, "What channel will it be on?"

56′ A student once threw a book at a mathematics instructor. "I wouldn't have done that," remarked another student. "Why not?" asked the culprit. "Because that's no way to treat a book," was the reply.

57′ Two friends were aboard ship and noticed a man leaning against the starboard rail.

"I'll bet he's a mathematics professor," ventured one.

"I know a mathematics professor when I see one," returned the other. "I'll bet you five dollars that he is not."

The other covered the bet and stepped up to the man. "I beg your pardon, but will you answer a question for us? Are you a mathematics professor?"

"No," he replied. "It's that I am seasick that makes me look the way I do."

58′ There was a lazy mathematics student who took up playing the trombone because it was the only instrument on which you can get anywhere by letting things slide.

59′ Blessed are they who go around in big circles for they shall be called big wheels.

60′ The evaluation inspectors were greatly impressed by the mathematics teacher of the school. Every time the teacher asked his class a question, all hands instantly shot up; and no matter on whom he called, he always received the correct answer.

Unknown to the inspectors, just prior to their visit, the teacher instructed his class, "When I ask a question, raise your right hand if you know the answer, raise your left hand if you don't."

61′   Our friends were the proud parents of their first baby boy. Thinking it only proper to name him after his father, Matthew, my nephew came up with an idea: "Why not call him the 'New Math'?"—RUTH J. ANDERSON

62′   A college freshman student, who had not done well in high school mathematics, decided to set aside three hours each evening for preparing his college algebra assignment. This strict schedule seemed to be working well until one evening he read the current assignment instructions: "Each of the problems 17 through 21 has an infinite number of solutions. Find them all."

63′   The mathematics professor was in the hospital for some delicate surgery. As he was undergoing a preliminary examination by the young surgeon, the doctor exclaimed, "Why, you're Professor Smith who taught mathematics at the University of Wisconsin. I had you for college algebra."

There was a long silence, and finally the professor asked, "How did you make out in the course?"

"Oh, you gave me a B plus," the doctor replied cheerfully.

"Thank goodness," murmured the professor from the examination table.

64′   One of the endearing things about mathematicians is the extent to which they will go to avoid doing any real work.
                                        —MATTHEW PORDAGE

65′   Yesterday a father heard a prophecy that the end of the world was coming next weekend. He repeated this to his son, who was cramming for a test. The boy's only answer was, "Good!"

66′   I once had a cross-eyed mathematics teacher who couldn't control her pupils.

67' Did you know that mathematicians are very symbol-minded people?

68' The mint should discontinue minting pennies—with inflation, they just don't make cents.

69' Bulletin descriptions of courses bear no relation to what the professors teach.
—Morris Kline
Why the Professor Can't Teach. St. Martin's Press, 1977, p. 10.

70' Universities hire professors the way some men choose wives—they want the ones the others will admire.
—Morris Kline
Why the Professor Can't Teach. St. Martin's Press, 1977, p. 92.

71' So far as the mere imparting of information is concerned, no university has had any justification for existence since the popularization of printing in the fifteenth century.
—Alfred North Whitehead
The Aims of Education.

72' Postulating properties has the advantage of theft over honest toil.—Bertrand Russell

73' At any mathematics conference, the two most interesting papers will be read at the same time.

74' A reflective man has learned that when he says "all" or "none" he means almost all or hardly any.
—W. Ward Fearnside and William B. Holther
Fallacy, the Counterfeit of Argument. Prentice-Hall, 1950, p. 12.

75' If your new theorem can be stated with great simplicity, then there will exist a pathological exception.
—Adrian Mathesis

76' All great theorems were discovered after midnight.
—Adrian Mathesis

77′   The greatest unsolved theorem in mathematics is why some people are better at it than others.—ADRIAN MATHESIS

78′   An optimistic gardener is one who believes that whatever goes down must come up.—FLOYD R. MILLER

79′   "Now, if you have that in your head," said the professor, who had just explained a mathematics theory to his students, "you have it in a nutshell."—THOMAS LAMANCE

80′   Epitaph: Here lies Eudoxus, who died of exhaustion.

81′   "Must we first take all these preliminary courses?" asked a mathematics student.
"There's only one endeavor in which one can start at the top, and that's digging a hole," replied the instructor.

82′   A beginning mathematics teacher put the following ad in the paper: "My services for hire—start haggling at about $17,000."
A superintendent responded with this message: "Bring own cigarettes and coffee—this may take time."

83′   *Superintendent:* "Do you want a $19,000 or a $17,000 job?"
Mathematics teacher: "What's the difference?"
Superintendent: "Well, we provide a bodyguard for the person who takes the $17,000 job."

84′   One man's Hermite is another man's Poisson.

85′   Referee's report: This paper contains much that is new and much that is true. Unfortunately, that which is true is not new and that which is new is not true.
—Heard at a mathematics meeting.

86′   Any program, by the time all bugs have been removed, is obsolete.

87' Computers are fantastic. In a few moments they can make a mistake so great that it would take many men many months to equal it.—M. Meacham

88' FORTRAN:

TRAN
TRAN
TRAN
TRAN

89' *Student:* Is Juan in the empty set?
Professor: No Juan is in the empty set.

90' Any school kid knows that George Washington was born in the year $1000 \sqrt{3}$.

SOME BITS AND TIPS
ON TEACHING MATHEMATICS

1' At the board you will make mistakes that no one even slightly familiar with the material could possibly make.
—Adrian Mathesis

2' It is customary to erase a chalkboard for the same reason we flush toilets.

3' It is also wise to erase a chalkboard because it destroys incriminating evidence.

4' At least once during a mathematics lecture you will say A, you will mean B, but the students will hear C, when all the time it should be D.

5' Teach to the problems, not to the text.—E. Kim Nebeuts

6' To state a theorem and then to show examples of it is literally to teach backwards.—E. Kim Nebeuts

7'   Teach no lesson before its time.

8'   Good mathematics books drive out bad or inferior mathematics books.

—BROTHER T. BRENDAN
The Mathematics Teacher, Feb. 1965.

9'   Never teach two things in the same lesson.

10'   A good preparation takes longer than the delivery.

—E. KIM NEBEUTS

11'   The words "figure" and "fictitious" both derive from the same Latin root, *fingere.* Beware!

—M. J. MORONEY
Facts from Figures, Penguin Books, 1977.

12'   "As I teach," said a mathematics teacher, "I try to keep in mind the axiom: 'There is nothing so unequal as equal treatment of unequals.' "

13'   Don't be a 2 × 4 mathematics teacher, one who always stays between the 2 covers of the textbook and within the 4 walls of the classroom.

14'   Some mathematics teachers talk in other people's sleep.

15'   A mathematics teacher should see to it that each of his students at some time achieves a marked success, and at some time gets an honest gauge of himself by a failure.

—WILLIAM H. BURHAM, adapted.

16'   Teaching is like an iceberg; seven-eighths of it is invisible from the surface.—ROBERT WEAVER, adapted.

17'   TV sets are invading the classroom. The coming ideal probably will be, not a professor on one end of a log and a student on the other, but a professor on one end of a coaxial cable and 50,000 students on the other.

18′   A teacher can never truly teach unless he is still learning himself. A lamp can never light another lamp unless it continues to burn its own flame. The teacher who has come to the end of his subject, who has no living traffic with his knowledge but merely repeats his lessons to his students, can only load their minds; he cannot quicken them.—Rabindranath Tagore

19′   Mathematics teachers are like the storage battery in an automobile—constantly discharging energy. Therefore, they need frequent recharging to forestall running dry.

20′   Because a man lives, it does not mean that he grows. At a university, a certain mathematics instructor was promoted to an assistant professorship. After the appointment was announced, another mathematics instructor approached his department chairman with a gnawing question.

"Why wasn't I given a promotion, too? After all, I've had twelve years teaching experience here."

"That's not quite the way I viewed it," came the reply. "In your case I felt that what you've had is *one* year's experience repeated twelve times."

21′   A great mathematics teacher is not one who imparts knowledge to his students, but one who awakens their interest in mathematics and makes them eager to pursue the subject for themselves. He is a spark plug, not a fuel pipe.

—M. J. Berrill, adapted.

22′   A mathematics teacher can light the lantern and put it in your hand, but you must walk into the dark.

—William H. Armstrong, adapted.

23′   In mathematics a train of thought is of little value unless it carries some freight with it.

24′   Mathematical food, like any other food, should be attractive and appetizing.

25′   Spoon feeding will only teach the shape of the spoon.

26' No person who thinks in terms of catching mice will ever catch lions.

27' Thinking is a habit like piano playing, not a process like eating and sleeping. The amount of thinking you can do at any time will depend primarily on the amount of thinking you have already done.

28' Learning mathematics is a lifelong process. Even if a genius could learn in college all the mathematics known, he could be out of college only a short time before the accumulation of new mathematics would make him a back number.

29' On the reverse side of Professor Doug Brumbaugh's personal card: If you were arrested for teaching, would there be enough evidence to convict you?

30' In problem solving, even if you are on the right track, you will get run over if you just sit there.

31' The secret of success in problem solving can be stated in nine words: Stick to it, Stick to it, Stick to it.

32' Nothing in problem solving can take the place of perseverance—talent alone will not, genius alone will not, education alone will not. It is perseverance that finally solves most problems. The slogan "press on" has solved and will solve many a mathematical problem. In other words, *stay*ability is more important than ability.

33' You can solve any problem if you have patience, claimed a noted mathematician. "Sometimes it takes time," he added.

To offer proof of his claim, he said: "You can carry water in a sieve—if you wait until it freezes."

34' Solving a difficult mathematics problem is much like cracking a hard nut. There is always a way to crack a hard nut, so long as you have the right kind of nutcracker.

35' A good mathematical idea must be hitched as well as hatched.

36' When you finally decide that a mathematics problem cannot be solved, just stand back and watch someone else solve it.

37' Making mathematics very abstract is like pouring hot water on delicate glasses; it can be done, but only after a warming period.

38' "Success," said the mathematics teacher to a failing student, "comes before work only in the dictionary."

39' Triumph is just *umph* added to *try*.

40' In mathematics you must treat ideas as though they were baby salmon. Throw a lot of them into the water. Only a few will survive, but they usually suffice.

41' The most brilliant mathematical ideas come in a flash, but the flash comes only after a lot of hard work. Nobody gets a big mathematical idea when he is not relaxed and nobody gets a big mathematical idea when he is relaxed all the time.

42' A child's progress in arithmetic might be improved if the ratio of praise to censure were at least 2 to 1.

43' The mathematics professor is the architect of her course, but the students must lay the bricks themselves.

44' If you copy anything out of one book, it is plagiarism. If you copy it out of two books, it is research. If you copy it out of six books, you are a professor.
> —From an address by Bishop FULTON J. SHEEN.

45' Intelligent ignorance is the first requirement in research.—CHARLES F. KETTERING

46'   Research is like saving. If postponed until needed, it is too late.

47'   The only way to avoid making mistakes in arithmetic is to gain experience, and the easiest way to gain experience in arithmetic is to make some mistakes.

48'   Last week I saw a man who had not made a mistake in four thousand years. He was a mummy in the British Museum.

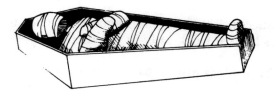

ILLUSTRATION FOR 48'

49'   Most mathematics problems can be solved with ordinary talent accompanied by extraordinary perseverance.

50'   Those who succeed in solving mathematics problems are not necessarily extraordinary; the rest of us haven't exerted ourselves enough.

51'   Advice to a student looking for a thesis topic in mathematics: "You don't have to climb the highest mountain to succeed. Still around are several molehills that haven't been scaled."

52'   There was more imagination in the head of Archimedes than in that of Homer.—VOLTAIRE

53'   To teach is to learn twice.

54'   Teachers affect eternity; they can never tell where their influence stops.

55'   Much of mathematics consists of attempts to replace difficult problems by easier ones having the same answers.

56'   Pure mathematics can be practically useful and applied mathematics can be artistically elegant.—PAUL HALMOS

57'   Euclid taught me that without assumptions there is no proof. Therefore, in any argument, examine the assumptions.
—E. T. BELL

58'   Mathematics consists in proving the most obvious things in the least obvious way.—GEORGE PÓLYA

59'   The art of doing mathematics consists in finding that special case which contains all the germs of generality.
—DAVID HILBERT

60'   Pure mathematics is on the whole distinctly more useful than applied mathematics. For what is useful above all is technique, and mathematical technique is taught mainly through pure mathematics.—G. H. HARDY

SOME LOGICAL AND SOME ILLOGICAL MOMENTS

1'   On the first day of school there were not enough seats for all the pupils. The teacher asked one little moppet, "Will you sit on a stool for the present?" The little girl went home at the end of the day, disillusioned, and told her parents, "I sat on that stool all morning and never did get a present."

2'   "Tommy, where are elephants found?" asked the teacher. "Elephants are so big that they hardly ever get lost," Tommy replied.

3'   "Tommy, how was iron discovered?" asked the teacher. "I heard Dad say they smelt it," Tommy replied.

4'   "Susie, what's your cat's name?" asked the teacher. "Ben Hur," Susie replied. "That's a funny name for a cat," said the teacher, "How did you happen to pick such a funny name for a cat?" "Well, we just called him Ben until he had kittens," was the reply.

5' "Susie, why in the fall do wild geese fly south?" asked the teacher. "Because it's too far to walk?" queried Susie.

6' Tommy announced to his parents that his reading class was to be divided into two divisions. "I'm in the top one," he said, "and the other is for backward readers. But we don't know who's going to be in the other one, because there's not a kid in the room who can read backward."

7' "Tommy, what's this low mark on your report card?" asked Tommy's dad. "Maybe it's the temperature of the school room," Tommy replied.

8' Susie came home from school with her January report card, which was anything but good. When her mother saw it, she cried out, "What happened this month?"
"Why, nothing unusual," answered Susie, "You oughta know— things are always marked down right after Christmas."

9' Susie, trying to explain the significance of her poor grades on the report card to her disgruntled dad, said, "Don't forget— we're studying all new stuff this year."

10' "I can't get that report card back for you," explained Tommy to his teacher. "You gave me an A in arithmetic and they're still mailing it to relatives."

11' A young teacher, imbued with the true spirit of her profession and aware of an excellent opportunity to emphasize citizenship through service, told the members of her class that we are here in this world to help others. The statement was well taken but one bright lad piped up, "What are the others here for?"

12'
Teacher: "What is your name, son?"
Small boy: "Jule, sir."

166

Teacher: "You shouldn't use a nickname. Your name must be
Julius. Next, what's your name?"
Second small boy: "Billious, sir."

13' In questioning the logic of her parents' actions, one girl
wanted to know why they insist she is too young and too little to
stay up late at night but the next morning tell her that she's too
old and too big to stay in bed.

14' Johnny could not restrain himself while the Sunday
school teacher told the story of Lot and his wife. When she explained the part in which Lot's wife looked back and turned into
a pillar of salt, little Johnny couldn't stand it any longer. Interrupting excitedly, he expostulated with fervor, "My mother looked
back once, as she was driving, and *she* turned into a fence post."

15' Epitaph: Here lies G. H. Hardy, with no apology.

16' The disappearance of dinosaurs from the earth has
been seriously accounted for by the fact that the animals were too
large to be admitted on Noah's Ark.

ILLUSTRATION FOR 16'

17' "Never waste household scraps," says an economy hint.
Agreed. Open the windows and let the neighbors hear.

18' There is positive proof that Americans are getting
stronger. In the early 1930s, when I was a teenage clerk in my

father's store, it took me two trips to carry two dollar's worth of groceries to a customer's car. Now my little five-year-old can easily carry that much in one load.

19'   The kindergarten teacher told of an animal lover who found a wounded dog by the side of the road, apparently hit by a passing car. He took the dog home wrapped in his coat and nursed it back to health. The teacher concluded by asking, "Do any of you children know of any such acts of kindness?"

Silence prevailed, then one little tyke said, "I didn't see this with my own eyes, but I heard Daddy say that he put his shirt on a horse and lost it."

20'   A Texas lad rushed home from kindergarten and insisted his mother buy him a set of pistols, holsters, and gun belt.

"Why, whatever for, dear?" his mother asked. "You're not going to tell me you need them for school?"

"Yes, I do," he asserted. "The teacher said tomorrow she's going to teach us to draw."

21'   From "Beetle Bailey": "What's that?" "An electric pencil sharpener." "Well, I'll be darned! I've never ever *seen* an electric pencil."

22'   A small boy was dolefully practicing his piano lesson when a salesman knocked on the door. "Son, is your mother home?" "What do you think?" answered the boy.

23'   A father of four boys came home to find them all engaged in something of a free-for-all. Addressing his remarks to the most aggressive of the four, he asked, "Butch, who started this?" "Well, it all started when Harold hit me back," exclaimed Butch.

24'   "Only an elephant or a whale gives birth to a creature whose weight is seventy kilograms or more. The president's weight is seventy-five kilograms. Therefore the president's mother was either an elephant or a whale."—STEFAN THEMERSON

25′ The dean quit his job at the university when he became a senator, thereby raising the average IQ of both the university and the Senate.

INDEX

This combined index contains all of the items from the original indexes in *Mathematical Circles Adieu* and *Return to Mathematical Circles*. References are to items, not to pages. Items from *Mathematical Circles Adieu* are preceded by an A. Items from *Return to Mathematical Circles* are preceded by an R. Thus A150 refers to Item 150° in *Mathematical Circles Adieu* and R150° refers to Item 150° in *Return to Mathematical Circles*.

A number followed by the letter *p* refers to the historical capsule just preceding the item of the given number (thus 275*p* refers to the historical capsule immediately preceding Item 275°).

Notations such as RMH 12, RTT 12, and RLM 12 refer, respectively, to Item 12° in Some Mathematical Humor in Minute Doses, Some Bits and Tips on Teaching Mathematics, and Some Logical and Illogical Moments, which are all sections of the Epilogue in *Return to Mathematical Circles*.